T0203126

# Radar Imaging of Airborne Targets

# Radar Imaging of Airborne Targets

## A Primer for Applied Mathematicians and Physicists

Brett Borden

*Naval Air Warfare Center Weapons Division*
*China Lake, California, USA*

**CRC Press**
Taylor & Francis Group
Boca Raton London New York

CRC Press is an imprint of the
Taylor & Francis Group, an **informa** business
A TAYLOR & FRANCIS BOOK

CRC Press
Taylor & Francis Group
6000 Broken Sound Parkway NW, Suite 300
Boca Raton, FL 33487-2742

First issued in paperback 2019

All text and illustrations in this work are in the public domain.
CRC Press is an imprint of Taylor & Francis Group, an Informa business

No claim to original U.S. Government works

ISBN-13: 978-0-415-25636-0 (hbk)
ISBN-13: 978-0-367-39185-0 (pbk)

---

**Library of Congress Cataloging-in-Publication Data**

---

Catalog record is available from the Library of Congress

---

**Visit the Taylor & Francis Web site at
http://www.taylorandfrancis.com**

**and the CRC Press Web site at
http://www.crcpress.com**

# Contents

# Preface

In the preface to volume 1 of the comprehensive *Radiation Laboratory Series*†, Louis Ridenour observes that: 'Radar is a very simple subject, and no special mathematical, physical, or engineering background is needed to read and understand [his] book.' While this was a relevant (and, undoubtedly correct) observation in 1947, the subject of radar—and, in particular, radar imaging—has evolved considerably since then. One consequence of this technological advancement is that the uninitiated present-day reader can be stymied by the profusion of jargon and implicit assumptions that frequently appear in the current literature, and the task of 'coming up to speed' in this area can be both challenging and time consuming.

My own experience has been that radar (and radar imaging) can, indeed, be considered a 'very simple subject'. Unfortunately, it is not really quite as simple as current practices seem to imply. The weak, non-interacting point scatterer target model, which is often too casually invoked by radar imaging developers, has had a long run, but it may be nearing the end of its predominance. Radar has undergone significant improvements in resolution and sensitivity in the past few decades, and radar systems are now being called upon to perform the increasingly demanding task of imaging the subscatterers that lie *within* the support of a 'traditional' target. As with any kind of imaging, optimal resolution and recognition will inevitably require accurate image models that, in this case, rely on accurate scattering analysis.

Very detailed electromagnetic scattering models, however, are not usually a part of the radar imaging story because they are, invariably, computation intensive and so are difficult to apply to high-speed radar environments. Rather, a 'middle ground' is usually taken where intermediate-level approximations are invoked as enhancements to the lowest-order 'weak scatterer' target model. This book is concerned with that 'middle ground'.

The targeted audience does not consist of the experienced radar engineers who would normally consult the 28 volumes of the 'Rad Lab' series. Instead, this discussion is intended to serve as an introduction to the subject, and has

† Ridenour L N 1947 *Radar System Engineering (Radiation Laboratory Series)* vol 1, editor-in-chief L N Ridenour (New York: McGraw-Hill).

been written in a manner which (it is hoped) will appeal to those physicists and applied mathematicians who are not approaching the topic of radar imaging from a formal radar background. The exposition concentrates on the direct scattering problem and the most common inverse scattering-based radar imaging techniques. The reader is assumed to have studied Classical Electromagnetics at the upper division level. Successful negotiation of an E&M course naturally presupposes a knowledge of Vector Calculus and Fourier Analysis at the level normally covered in an upper division 'Applied Analysis' course. Some knowledge of Probability and Statistics, as well as Complex Variable theory, will also be helpful but is not essential. In short, the following discussion should be readily accessible to graduate-level students of applied mathematics and the physical sciences.

This work was supported by the Office of Naval Research and special appreciation is extended to Bill Miceli who first suggested that 'an overview' be written. The final form of the manuscript owes much to the careful reading by Carey Schwartz and the many useful suggestions that he provided. In addition, Alan Van Nevel, Gerry Kaiser and Sam Ghaleb voluntarily proofed various portions of the text, the equations and the conceptual development—an effort that improved both correctness and clarity, and for which I am very grateful. Finally, I'm also indebted to Arje Nachman of the Air Force Office of Scientific Research for some six years of encouragement and for a meticulous reading of the manuscript which discovered numerous errors that the others missed.

**Brett Borden**                                    *China Lake, California*
                                                    January 1999

# 1

---

# Introduction

Sensor systems that can detect, locate and identify targets at great distances and in all kinds of weather have well-recognized utility. To date, the only systems that can perform the long-range and all-weather detection and location functions are radar-based. This is because the wavelength of radar signals makes them relatively unaffected by atmospheric and weather-induced attenuation. But image resolution also depends upon signal wavelength, and an inconvenient side effect of this long-wavelength property is that radar-based target identification schemes will be fundamentally less effective than many of their optical-based imaging counterparts. In an effort to get around this limitation, a number of radar-specific methods for automatic target recognition (ATR) have been proposed. Some of these techniques include jet engine modulation, target scattering resonances, target surface vibration effects, nonlinear joint-contact effects, high range resolution (HRR) imaging systems and synthetic and inverse synthetic aperture radar techniques (SAR and ISAR). In those military environments for which airborne systems are required to identify/classify aircraft targets, contemporary research and development efforts have concentrated on HRR and ISAR imaging methods.

Because of the historical evolution of radar technology, most current HRR and ISAR imaging algorithms appear to have been built upon a foundation which views targets as being composed of simple, non-interacting point scatterers. This scattering model was originally devised for the problem of detection of single aircraft but, as resolution began to improve, it was also seen to be accurate and useful for echo-location of multiple aircraft targets. As radar technology was further refined to the point where the *components* of a single target could be resolved, it was natural to try to interpret an aircraft target as a collection of (non-interacting) multiple subtargets—and, for the most part, this has worked pretty well. In recent years, however, radar resolution and sensitivity have reached a level that the limitations of this old model are often apparent. HRR and ISAR images created under the weak scatterer approximation may display artifacts which are decidedly not point-like and can confound the target identification process. Despite this limitation, current approaches continue to use the weak scatterer approximation as a first-order expedient and append 'correction terms'

1

to the imaging algorithm to account for any image artifacts that might result. Motivated by 'ease of interpretation' arguments, these correction terms may be poorly defined and sometimes exist only as an *ad hoc* collection of informal filters used to weight the importance of the image elements.     Such filters are 'artistically' applied by savant practitioners in ways that may defy formal explanation.

## 1.1   BRIEF HISTORY OF RADAR

The name 'radar'—which is an acronym for the phrase 'RAdio Detection And Ranging'—was coined in late 1940, but the earliest radar-like devices pre-date this appellation by almost four decades.  (As a palindrome, this name also encapsulates the basic send/echo idea behind these measurement systems.)  Of course, the principles underlying radar can be formally traced back to Maxwell (and the theoretical foundation of all classical electromagnetic phenomena), but this seems to imply an idealized sequential development from theory to experiment that does not really apply to the radar story.  Rather, it was probably Heinrich Hertz who started the idea of radio echo-location in the last quarter of the nineteenth century by *experimentally* demonstrating that radio waves could be reflected from objects.

Devotees of Nikola Tesla sometimes attribute the 'invention' of radar to his legendary year of 1899 in Colorado Springs.  It is more realistic to say that Tesla's actual contribution appears to be only a conceptual description that was probably unknown by other remote-sensing investigators until after the first radar sets were being refined.  The distinction of 'first electromagnetic echo-location device' is generally awarded to a ship anti-collision system developed by Christian Hülsmeyer in 1904.  Hülsmeyer's invention operated at a frequency of about 700 MHz and was successfully demonstrated to the German Navy and a Dutch shipping company.  Because of its limited range, however, it was never put into service.

Radar development began in earnest after Taylor and Young (who were studying VHF radio propagation for the US Navy in 1922) observed reflection phenomena from buildings, trees and other objects.  These results were based on 60 MHz continuous-wave measurements but, at about the same time, radio *pulse* echo measurements of the altitude of the ionosphere were also (independently) performed.  In 1930, Young—this time with Hyland—observed antenna pattern variations that were due to passing aircraft.  This event initiated the first significant radar research effort for the United States (under the direction of the Naval Research Laboratory).

In the 1930s, Great Britain and Germany also began their own radar programs. The British program appears to have begun as a fallback effort when the original idea of an electromagnetic 'death ray' was shown to be unfeasible as a method for air defense. The German effort, which was initially applied to ship detection, quickly included aircraft echo-location as an important goal.

The pre-war British radar aircraft detection effort resulted in the famous Chain Home early warning system. This widespread collection of radar transmitters and receivers operated (typically) at about 22–50 MHz and used horizontal polarization which lessened sea backscatter and was believed to optimize aircraft return (the wingspan of the target aircraft was modeled as a half-wave dipole with horizontal orientation).

British and German radar systems were also used for fire control, aircraft navigation and, late in the decade, the British succeeded in developing an airborne radar. These radar systems were operated at frequencies of from tens to hundreds of MHz. Higher frequencies, which could increase antenna directivity and allow for smaller antenna sizes, were not used because efficient methods for their generation were unknown—the klystron tube, which was developed in the late 1930s as a microwave source, lacked the necessary power for radar applications. The big breakthrough came with the British development of the cavity magnetron at the end of the decade.

In 1940 the British began a collaborative research effort with the United States and, at this time, US researchers were informed of the existence of the cavity magnetron as a means for producing high-power microwaves. A new research center was established at the Radiation Laboratory at MIT and a usable microwave tracking radar, capable of being mounted in an aircraft, followed in less than six months.

Other microwave radar systems were developed for artillery fire-control and for improved bombing accuracy. In fact, it was microwave radar components captured from downed British and American aircraft that led Germany to adopt the cavity magnetron and begin manufacturing their own microwave radars—although these came too late to have a significant wartime impact. By the end of the war, radar frequencies of 10 GHz were common.

Radar developments during the war were transferred to peacetime applications afterward. Notable among these were navigation aids for commercial ships and aircraft as well as shipborne collision avoidance systems. It was the Cold War and Korean conflict, however, that re-invigorated many of the radar research efforts that were downsized after World War II.

Improvements in radar sensitivity and the application of frequency shift (Doppler shift) measurements for determining target speed—which were never really used during the war—enabled radars to meet the more demanding requirements imposed by high-speed jet aircraft. In addition, radar sets became small enough to be placed within guided missiles. Advances in signal processing algorithms allowed for the development of increased effective range and improved clutter rejection techniques. Significantly, the tendency toward computer control of radars and computer processing of radar data has become something of a 'theme' in modern radar research and development [1–4].

## 1.2  CONTEMPORARY ISSUES IN RADAR IMAGING

The advent of solid-state electronics has had a significant impact on radar development, and progress in the 1960s was prodigious and diverse. Within the context of radar-based imaging, several important accomplishments were made at this time: high-resolution SAR was introduced in the middle of this decade; and a system with range resolution of about 1 m was developed a little later. Also notable was the development of a 95 GHz radar and the first extensive studies of radar-based target recognition [5, 6].

The application of 'inverse scattering' ideas to radar target identification appears to have begun in the mid 1960s and grew to an important area of research in the 1970s [7, 8]. This growth was concurrent with the rise of minicomputers and microprocessors. The 1970s also saw the development of high-range-resolution monopulse radar systems and active missile guidance systems.

The trend of higher frequencies and greater bandwidth—which, as we shall see, are independently required for effective radar imaging—was largely responsible for the application of SAR-like ideas to airborne target imaging in the late 1970s. Of course, the computational issues were found to be considerable and it is doubtful that frequency and bandwidth factors alone could have justified many of the original ISAR *development* efforts. Rather, it was the parallel advancements in computer technology that were also occurring at this time that truly enabled practicable radar imaging of airborne targets to be contemplated. And, as predicted in [1], digital signal processing techniques have dominated 'radar' development ever since.

The (correct) prediction of the importance of digital processing techniques to radar was pretty easy to make and was the majority opinion in the radar community during the late 1970s. At the same time another prediction was widely held but has not come true; namely, that radar-based target recognition systems would be commonplace by the year 2000. The reasons behind this failure—in spite of several decades of intense research—can, in many cases, be traced to failed intuition and 'wishful thinking'. This intuition appears to have been based on a weak, non-interacting scattering model approximation that was developed for multiple aircraft targets, and not for multiple components of a single aircraft target. A more careful analysis has shown that there was really never any reason to believe that the approximation would hold in the latter case and current research efforts have concentrated on devising signal processing methods to 'get around' the limitations associated with this incorrect assumption.

In recent years it has become quite apparent that the problems associated with radar target identification efforts will not vanish with the development of more sensitive radar receivers or increased signal-to-noise levels. In addition, it has (more slowly) been realized that greater amounts of data—or even additional 'kinds' of radar data, such as added polarizations or greatly extended bandwidths—will all suffer from the same basic limitations affiliated with incorrect model assumptions. In the face of these problems it is important to ask how (and if) the complications associated with radar-based ATR can be surmounted.

## 1.3   OVERVIEW

Due, in part, to long-standing traditions in radar development, advanced algorithm design must often accommodate limitations built into older systems which were created to exploit the weak scatterer model. Combined with real-time processing requirements and data restrictions associated with airborne encounters, the problem of fitting a new algorithm to an existing system can be quite challenging. Below, we will describe some of the problems that must be addressed before radar-based airborne ATR can become a reliable tool.

This discussion is, principally, about radar techniques and we start with a brief description of radar measurement systems and what they actually measure (chapter 2). The approach taken is not complete and is intended to mathematically describe radar transmission and reception without appealing to block diagrams or device physics—the uninterested reader can briefly scan this chapter for highlights without loss of continuity. The scattering models developed in chapter 3 are important to the later discussions, however, since we will argue that incorrect scattering model assumptions are largely responsible for radar image errors. Consequently, we will take some care in establishing the 'standard' radar image model from Maxwell's equations and detail the approximations that are often taken for granted by many radar developers. Chapters 2 and 3 also serve to establish our basic notation and coordinate conventions.

Because of the attention they are currently receiving, we will concentrate on HRR and ISAR techniques. Chapter 4 introduces the simplest type of radar image—the 'range profile'. Owing to the simplicity of range profiles, the bulk of this chapter is actually devoted to problems associated with first-kind integral equations (which describe range profiles). Some of the more common resolution enhancement methods proposed for HRR target recognition will also be explained in this chapter. Standard ISAR methods are discussed in chapter 5. This kind of two-dimensional radar imaging is also described by a first-kind integral equation but, in practice, ISAR imaging is much more difficult than HRR and so this chapter will be almost entirely devoted to implementation considerations of the usual ISAR imaging algorithm. (The last section of chapter 5 introduces a generalization of traditional two-dimensional radar imaging and has been included as a transition to section 8.4.)

It is not really until chapter 6 that we explain some of the difficulties in applying these results to the issue of target classification and recognition and describe various approaches that have been taken toward addressing them. As an important example of the non-weak scatterer exception to the standard scattering model, we will examine the problem of duct/cavity-induced image artifacts in somewhat greater detail. Chapter 7 summarizes the straightforward ideas behind three-dimensional imaging based on angle-of-arrival measurements. Finally, and for a semblance of 'completeness', we will outline various other (mostly non-imaging) radar-based target classification and identification methods that have been proposed and/or employed (chapter 8).

## REFERENCES

[1]   Walsh T E 1978 Military radar systems: history, current position, and future forecast *Microwave J.* **21**(11) 87

[2]   Swords S S 1986 *Technical History of the Beginnings of Radar* (London: Peregrinus)

[3]   Bowen E G 1987 *Radar Days* (Bristol: Adam Hilger)

[4]   Fisher D E 1988 *A Race on the Edge of Time: Radar—the Decisive Weapon of World War II* (New York: McGraw-Hill)

[5]   Rosenbaum-Raz S 1976 On scatterer reconstruction from far-field data *IEEE Trans. Antennas Propag.* **24** 66

[6]   Lewis R M 1969 Physical optics inverse diffraction *IEEE Trans. Antennas Propag.* **17** 308; correction 1970 *IEEE Trans. Antennas Propag.* **18** 194

[7]   Baltes H P 1980 *Inverse Scattering Problems in Optics, Topics in Current Physics* ed H P Baltes (New York: Springer)

[8]   Boerner W-M 1985 *Inverse Methods in Electromagnetic Imaging (NATO ASI Series C: Mathematical and Physical Sciences)* vol 143, ed W-M Boerner *et al* (Dordrecht: Reidel)

# 2

---

# Radar Fundamentals

The original purpose of radar echo 'target'-location remains the principal function of modern radar systems and their design typically results in a long list of engineering compromises that are chosen to solve the combined problems of detecting targets at maximum range and allowing for the application of straightforward signal processing algorithms. Consequently, radar systems have evolved over many decades to take advantage of relevant improvements in electronic technology. From the perspective of a mathematical analysis of radar/target interactions, this collection of engineering compromises means that a detailed understanding of the behavior of radar systems will not be easily achieved—such information traditionally fills volumes and defies attempts at simple and abbreviated explanations.

Fortunately, we will not require a detailed understanding of the inner workings of radar systems to develop the theory of radar imaging. In particular, we will only need to know that radar waveforms can be represented as complex-valued solutions to the wave equation and that radar systems are able to measure these signals (in both amplitude and phase) as functions of time. Nevertheless, it will be useful to gain a basic understanding of how a radar system actually goes about accomplishing this measurement task because it helps in our understanding of the limitations inherent to these radar data. The present chapter offers a greatly simplified overview of radar signals and the (hardware-based) techniques used to process them. It affords the reader an opportunity to become familiar with some of the terminology used to describe radar signals and to appreciate some of the engineering compromises that have been made.

Three basic issues will be addressed. The first is concerned with the kind of signals that a radar uses to generate the waveforms it transmits. These signals are real-valued functions of time but can be considered to be the real *parts* of complex-valued functions, and the second issue concerns this practice. Once defined, the imaginary part of the complex-valued signal will determine the phase of the waveform and, as we shall see in later chapters, considerable information about the target can be extracted from this phase. Physically, radar signals are simply time-varying voltages within the radar system and, in principle, could be almost any causal and finite-energy function. The waveforms that these signals generate, however, must satisfy the wave equation and the radiation condition. The echo waveform—that part of the transmitted wave scattered from a radar target—will also satisfy the wave equation and radiation condition and these properties imbue the radar echo with important target-specific information. The third issue that we will examine, then, is the means by which a radar system uses measurements of the echo waveform to estimate the range, speed and bearing of the target. The discussion is by no means complete and the interested reader is encouraged to consult the references for further details.

## 2.1   RADAR SIGNALS

Ultimately, radar waveforms are measured as time-domain voltages $s(t)$ and are subject to ordinary signal processing methods. These methods can be applied in either the time domain or the frequency domain—each approach being related through the Fourier transform:

$$S(\omega) = \mathcal{F}\{s\}(\omega) = \int_{\mathbb{R}} s(t')e^{-i\omega t'}\,dt' \tag{2.1-1}$$

with corresponding inverse transform

$$s(t) = \mathcal{F}^{-1}\{S\}(t) = \frac{1}{2\pi}\int_{\mathbb{R}} S(\omega')e^{i\omega't}\,d\omega' . \tag{2.1-2}$$

Let $s^*(t)$ denote the complex conjugate of $s(t)$. The autocorrelation of $s(t)$ is given by

$$\mathcal{A}(t) = \int_{\mathbb{R}} s(t')s^*(t'+t)\,dt' . \tag{2.1-3}$$

The autocorrelation function is especially useful because of its relationship to the *power spectrum* $\mathcal{P}(\omega)$ of $s(t)$. For our purposes, the power spectrum can be defined by

$$\mathcal{P}(\omega) = \int_{\mathbb{R}} \mathcal{A}(t')e^{-i\omega t'}\,dt' = |S(\omega)|^2 . \tag{2.1-4}$$

$\mathcal{P}(\omega)$ represents the fraction of a signal's energy lying between the (angular) frequencies $\omega$ and $\omega+d\omega$ in the Fourier domain. When $s(t)$ is real-valued, $\mathcal{P}(\omega)$ will be an even function of $\omega$ and we must exercise some care in crediting this energy density. If we assume that $s(t)$ is normalized so that its total energy $\int_{\mathbb{R}} |s(t')|^2\,dt' = 1$ (which implies $\int_{\mathbb{R}} \mathcal{P}(\omega')\,d\omega' = 2\pi$), then the average frequency and bandwidth can be defined in terms of the first and second moments of *one side* of $\mathcal{P}(\omega)$ by

$$\bar{\omega} = \frac{1}{\pi}\int_{\mathbb{R}^+} \omega'\mathcal{P}(\omega')\,d\omega'$$

$$\beta = \left(\frac{1}{\pi}\int_{\mathbb{R}^+} (\omega'-\bar{\omega})^2\mathcal{P}(\omega')\,d\omega'\right)^{1/2} . \tag{2.1-5}$$

(We will explain these limits of integration more carefully in the following section.)

It is the bandwidth that is important to problems in radar target imaging since there is a relationship between time-domain resolution and frequency-domain bandwidth. Using the properties of Fourier transforms, it is straightforward to show that $\beta$ can be expressed directly in terms of $s(t)$ as

$$\beta = \left( \int_{\mathbb{R}} \left| \frac{ds(t')}{dt'} \right|^2 \, dt' - \overline{\omega}^2 \right)^{1/2} . \tag{2.1-6}$$

Radar systems will transmit and receive *real-valued* waveforms, and the (real-valued) signal $p(t)$ that radar engineers choose is conventionally written as

$$p(t) = a(t) \cos(\phi(t) + \omega_0 t) \tag{2.1-7}$$

where $\omega_0$ is the 'carrier' frequency. The function $a(t)$ is known as the *envelope* and $\phi(t)$ is the *phase modulation*. The continuous signal $\cos \omega_0 t$ is called the *carrier signal*. The envelope and phase modulation functions are (usually) chosen to be slowly varying functions relative to the carrier signal and, consequently, the Fourier transform will have frequency components concentrated around the carrier frequency. This *narrow band* assumption will be very important for the remainder of this discussion and, in practice (where typically $\omega_0 \sim 10^{10}$ Hz), 'narrow band' still affords us considerable freedom in signal design. (We will consider generalizations to the narrow band assumption when we examine 'wide band' waveforms in section 8.4.) The carrier frequency is often set to the average frequency and we will always take $\omega_0 = \overline{\omega}$ (unless otherwise stated). It will also be convenient for us to write

$$\Phi(t) = \phi(t) + \omega_0 t . \tag{2.1-8}$$

In section 2.5 we will show how target location (in range) can be measured using short radar pulses and it is easy to see how these pulses can be generated using equation (2.1-7) by setting, for example, $a(t)$ to a sum of time-translated rect$(t)$ functions and $\phi(t) = 0$. (Recall that rect$(t) = 1$ over a support $t \in (-1/2, 1/2)$ and rect$(t) = 0$ otherwise.) The total energy that can be put into this pulse will be limited to the product of the peak power of the radar transmitter and the duration of the pulse and, consequently, the shorter the pulse, the less energy can be 'put onto' the target.

Frequency modulated carrier waves are typically used by radar engineers as a practical method for increasing the amount of energy in a short pulse without having to increase the transmitter's peak power. Substituting equations (2.1-7) and (2.1-8) into (2.1-6) yields

$$\beta^2 = \frac{1}{2} \int_{\mathbb{R}} \left[ \left| \frac{da(t')}{dt'} \right|^2 + |a(t')|^2 \left( \frac{d\Phi(t')}{dt'} \right)^2 \right] dt' - \omega_0^2$$

$$+ \frac{1}{2} \int_{\mathbb{R}} \left[ \left| \frac{da(t')}{dt'} \right|^2 - |a(t')|^2 \left( \frac{d\Phi(t')}{dt'} \right)^2 \right] \cos 2\Phi(t') \, dt' \quad (2.1\text{--}9)$$

$$- \int_{\mathbb{R}} a(t') \frac{da(t')}{dt'} \frac{d\Phi(t')}{dt'} \sin 2\Phi(t') \, dt'$$

(where $\int_{\mathbb{R}} |p(t')|^2 \, dt' = 1 \Rightarrow \int_{\mathbb{R}} |a(t')|^2 \, dt' = 1$). When $a(t)$ and $\phi(t)$ are slowly varying (with respect to $\omega_0 t$), the integrands containing $\cos 2(\phi + \omega_0 t)$ and $\sin 2(\phi + \omega_0 t)$ will (under integration) make a vanishingly small contribution to $\beta^2$. Consequently, under the narrow band assumption, equation (2.1–9) becomes

$$\beta^2 \approx \frac{1}{2} \int_{\mathbb{R}} \left| \frac{da(t')}{dt'} \right|^2 dt' + \frac{1}{2} \int_{\mathbb{R}} |a(t')|^2 \left( \frac{d\Phi(t')}{dt'} \right)^2 dt' - \omega_0^2. \quad (2.1\text{--}10)$$

From this, it is easy to see that a *nonlinear* $\phi(t)$ will always increase the bandwidth in comparison with signals that are modulated in amplitude only. (Linear $\phi$ would simply change the value of $\omega_0 = \bar{\omega}$.) When the signal is phase modulated by a nonlinear $\phi$, the signal is said to be 'pulse-compressed'. The 'pulse-compression ratio' is defined as the product of the pulse spectral bandwidth $\beta$ and the uncompressed pulsewidth $\tau$ and values of $\beta\tau \sim 100$ are common.

As a time-domain voltage in the circuits of the radar system, the transmission signal is used to create a transmitted waveform. Conversely, the echo waveform induces a reception signal in the radar receiver. Since they travel through space, all radar *waveforms* $w$ must be solutions to the wave equation. If $c$ denotes the propagation speed of the waveform, then the one-dimensional wave equation is

$$\frac{\partial^2 w(y,t)}{\partial y^2} - \frac{1}{c^2} \frac{\partial^2 w(y,t)}{\partial t^2}$$

$$= \left( \frac{\partial}{\partial y} + \frac{1}{c} \frac{\partial}{\partial t} \right) \left( \frac{\partial}{\partial y} - \frac{1}{c} \frac{\partial}{\partial t} \right) w(y,t) = 0. \quad (2.1\text{--}11)$$

Under the change of variables $u = y/c + t$ and $v = y/c - t$, equation (2.1–11) can be written as

$$\frac{\partial^2 w(y,t)}{\partial u \partial v} = 0 \quad (2.1\text{--}12)$$

which has a straightforward solution

$$w(y,t) = A \, f_1(y/c - t) + B \, f_2(y/c + t) \quad (2.1\text{--}13)$$

in terms of any twice-differentiable functions $f_1$ and $f_2$.

We will choose the convention that a Galilean observer traveling with the *transmitted* wave (in a positive direction along the $y$-axis) should always see a fixed signal voltage. In the frame of the radar, this convention implies that solutions will be of the form

$$w(y, t) = \begin{cases} s(y/c - t), & \text{for the transmitted wave} \\ s(y/c + t), & \text{for the echo wave.} \end{cases} \quad (2.1\text{--}14)$$

## 2.2  RADIATION CONDITION

For reasons that will be made clear in section 2.5.2, radar engineers are especially interested in examining radar signals in the frequency domain. Of course, frequency-sensitive radar systems are necessarily band-limited and, consequently, so will the radar data that they collect. Band-limited data have the interesting property of being easily extended to complex-valued functions which are analytic in (regions of) the complex time plane and, even though analytic signals are 'unphysical' (in the sense that they are not of finite duration), this analyticity has some useful benefits.

The radar transmitted pulse and echo are real-valued but, for signal processing purposes, are sometimes more conveniently treated as the real parts of complex-valued waveforms $w(y, t) = p(y/c \pm t) + iq(y/c \pm t)$. Generally, there is no unique way to assign the imaginary part of such a signal. If, however, we require the resulting complex-valued waveform to be analytic in half of the complex time plane, then its real and imaginary parts must form a Hilbert transform pair and we can write

$$q(t) = \mathcal{H}\{p\}(t) = \frac{1}{\pi} \fint_{\mathbb{R}} \frac{p(t')}{t - t'} \, dt' \quad (2.2\text{--}1)$$

where $\fint (\cdots)$ denotes the principal value of the integral.

The Hilbert transform can be expressed in terms of the Fourier transform as $\mathcal{H}\{p\}(t) = -i\mathcal{F}^{-1}\{\text{sgn}(\omega)\,\mathcal{F}\{p\}(\omega)\}(t)$, where $\text{sgn}(\omega)$ denotes the sign of $\omega$. Consequently, it is easy to see that $s(t) = p(t) + i\mathcal{H}\{p\}(t)$ will have non-zero spectral components only when $\omega \in \mathbb{R}^+$. (We shall return to this important point at the end of section 8.4.)

Because of equation (2.1–4), the requirement (2.2–1) will express the imaginary part of $w(y, t)$ in terms of an integral of its real part *over all time* and so is impossible to implement in practice. The approach that radar engineers have used to address this problem relies on the observation that $\cos at$ and $\sin at$ are Hilbert transform pairs. Consequently—under the narrow band assumption—we can approximate the Hilbert transform of the signal (2.1–7) as†

$$q(t) \approx a(t) \sin (\phi(t) + \omega_0 t) . \quad (2.2\text{--}2)$$

---

† Actually, in this case the narrow band assumption can be relaxed to the requirement that the spectrum of $a(t) \exp(i\Phi(t))$ should be single-sided [7].

This is known as the *quadrature* component of $w$ and can be readily implemented by a $\pi/2$ phaseshifter. (The corresponding $p(t)$ is known as the *in-phase* component.) This quadrature model allows us to write

$$s(t) = p(t) + iq(t) = a(t)e^{i\Phi(t)} \tag{2.2-3}$$

and, within the radar literature, this definition is so pervasive that the term 'quadrature' is often dropped.

The wave equation in $\mathbb{R}^3$ is an easy generalization of equation (2.1–11) although it is often more useful to consider its reduced forms. When the radar wave is considered to be of the form of equation (2.2–3) the wave equation becomes

$$\nabla^2 w(x, t) + \frac{1}{c^2} \left( \frac{d\Phi(t)}{dt} \right)^2 w(x, t) = 0 . \tag{2.2-4}$$

This equation is approximate and relies on the narrow band assumption to discard the other time derivative terms as insignificantly small. When equation (2.2–4) is accurate we can define the 'instantaneous frequency' in terms of $\Phi(t) = \phi(t) + \omega_0 t$ as

$$\omega_i = \frac{d\Phi}{dt} . \tag{2.2-5}$$

(Note, however, that $\Phi$ will generally be defined through the Hilbert transform relationship of equation (2.2–1) and so $\omega_i$ will not really be 'instantaneous'.)

Each of the Fourier components $W(x, \omega)$ of $w(x, t)$, defined by equation (2.1–1), is easily seen to be a wave of the form of equation (2.2–3) and satisfies (2.2–4) exactly (with $d\Phi/dt = \omega$, the component frequency). In this special case, equation (2.2–4) is independent of $t$ and is known as the Helmholtz equation. (It is important to observe, however, that $\omega_i$ should not be confused with the Fourier component frequency $\omega$—they are generally not the same and we shall return to this point in section 2.7.)

Uniqueness of the solutions to the Helmholtz equation will be assured if they are required to obey the boundary conditions of a wave radiating outward from a radar antenna (source). These boundary conditions require that the solutions decay to zero at a rate of $1/R$ as $R \to \infty$ and that they be outgoing. The first requirement is

$$W(x, \omega) = O(R^{-1}) , \qquad R \to \infty . \tag{2.2-6}$$

The second condition is determined by differentiation with respect to $R$. It requires

$$\frac{\partial W}{\partial r} - \frac{i\omega}{c} W = o(R^{-1}) , \qquad R \to \infty . \tag{2.2-7}$$

(Equation (2.2–7) is known as the Sommerfeld radiation condition.)

When $\omega_i$ is slowly varying, the 'outgoing' Green's function (elemental solution) to equation (2.2–4) is

$$G(x, x') = \frac{e^{ikR}}{4\pi R} \tag{2.2-8}$$

where $R = |x - x'|$ and $k = \omega_i/c$.

## 2.3 THE RADAR EQUATION

The so-called 'radar equation' relates the power density of the echo waveform to the power density transmitted by the radar and the range (distance) to the target. This will be the ideal radar receiver power density and so the radar equation can be used to estimate the radar's maximum range performance.

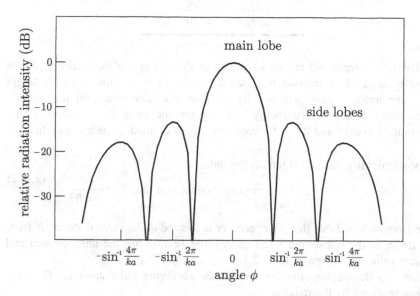

**Figure 2.1** Antenna radiation pattern for a one-dimensional uniform aperture distribution. The antenna is of length $a$ centered on the origin and the radiation pattern obeys $(\frac{1}{2}ka \sin\phi)^{-1} \sin(\frac{1}{2}ka \sin\phi)$.

Let $P_T$ denote the power density transmitted by the radar antenna. If a target is located at a distance $R$ from this transmitter, then the boundary condition of equation (2.2–6) requires the power density at a target to be reduced to the value

$$\text{power density at target} = \frac{P_T g}{4\pi R^2} \qquad (2.3–1)$$

where $g$ is the *gain* of the transmitting antenna. The factor $g$ accounts for an antenna's directivity: $4\pi g = 1$ for an isotropic antenna while non-isotropic antennas will be designed so that, typically, $4\pi g > 1$ in the direction of the target and $4\pi g < 1$ in other directions. A detailed description of antenna gain is beyond the scope of this discussion but it is useful to see a 'characteristic' plot of an antenna radiation pattern displayed as a function of off-axis angle (figure 2.1).

**Table 2.1** Typical radar cross section values.

| Target | $\varsigma$ (m$^2$) |
|---|---|
| House fly | $10^{-5}$ |
| Bird | $10^{-2}$ |
| Air-to-air missile | $10^{-1}$ |
| Jet fighter | $10^{1}$ |
| Jet airliner | $10^{2}$ |

The radar target will reflect a fraction of this energy in the direction of the radar receiver. This fraction is called the radar cross section $\varsigma$ of the target and, like antenna gain, will usually be directive (depending on the target's orientation). This reflected energy will undergo the same $R^{-2}$ decrease as the transmitted energy and the radar cross section is defined in such a way that

power density of echo signal at the radar

$$= \text{power density at target} \times \frac{\varsigma}{4\pi R^2} . \qquad (2.3\text{--}2)$$

(We have assumed that the radar receiver is located at the same distance $R$ from the target as the transmitter.) The radar cross section has the units of area and typical values are given in table 2.1.

Let $A_{\text{eff}}$ denote the effective area of the receiving radar antenna. Then the power received by the radar is

$$P_{\text{r}} = \frac{P_{\text{T}} g}{4\pi R^2} \times \frac{\varsigma}{4\pi R^2} \times A_{\text{eff}} = \frac{P_{\text{T}} g A_{\text{eff}} \varsigma}{(4\pi)^2 R^4} . \qquad (2.3\text{--}3)$$

This is the 'radar equation': the $R^{-4}$ law that relates target detectability to radar power. If $P_{\text{min}}$ is the minimum signal power detectable by the radar, then the maximum radar detection range will be

$$R_{\text{max}} = \left( \frac{P_{\text{T}} g A_{\text{eff}} \varsigma}{(4\pi)^2 P_{\text{min}}} \right)^{1/4} . \qquad (2.3\text{--}4)$$

In practice, this simple expression does not completely describe the behavior of target detectability and, usually, $R_{\text{max}}$ will be less than the value given by equation (2.3–4). There are a variety of factors that will also affect the estimate $R_{\text{max}}$ and these are usually addressed in the design of actual radar systems. Most of these considerations are beyond the scope of this discussion (cf [3]) but the issue of atmospheric attenuation bears some mention since it is often used to determine the frequency band that a radar system will employ.

## 2.4   ATMOSPHERIC 'WINDOWS'

In developing equation (2.3–4) we have assumed that the radar waveform propagates without attenuation. Because of atmospheric absorption, however, a portion of the waveform energy will be lost during transmission. This energy loss is strongly affected by the amount of water vapor in the atmosphere at any given time and will also depend on the frequency of the radar waveform in question.

Figure 2.2 is a plot of atmospheric attenuation as a function of radar frequency under 'clear atmosphere' conditions (1 atmosphere of pressure, temperature at 293 K, and oxygen and water vapor at 7.5 g m$^{-3}$).

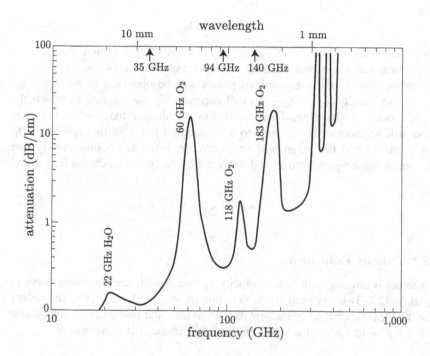

**Figure 2.2**   Attenuation of electromagnetic energy propagating in the atmosphere (based on [6]).

Figure 2.2 displays the existence of several 'atmospheric windows'— frequency bands for which radar waveforms suffer (relatively) small energy loss. These windows are largely responsible for the interest in 10 GHz, 35 GHz and 94 GHz as center frequencies 'useful' to radar system design.

## 2.5  RADAR DATA

For convenience, we will consider the target to be a single point scatterer. This approximation will be 'justified' in the next chapter as being appropriate to high-frequency waveforms scattered from weak, far-field scatterers and is the traditional model used for developing the ambiguity function.

### 2.5.1  Range Estimation

A radar estimates a target's 'range' (the distance between the radar and the target) by transmitting a (typically) short pulse and measuring the delay time of the echo pulse. If $t_d$ is this delay time and $v$ is the radial velocity of the target then

$$R = \tfrac{1}{2}(c + v)\, t_d \qquad (2.5\text{--}1)$$

where the factor of $\tfrac{1}{2}$ accounts for the two-way pulse propagation distance.

Practical radar systems transmit multiple pulses at times $t_{t_1}$, $t_{t_2}$, $t_{t_3}$, ..... The echoes due to these transmitted pulses will be observed at times $t_{e_1}$, $t_{e_2}$, $t_{e_3}$, ... and, ideally, $t_{d_n} = t_{e_n} - t_{t_n}$ will depend only on the rate at which the target's range is changing. There is always the danger that a detected echo pulse will be incorrectly assigned to a transmitted pulse so that $t_d = t_{e_3} - t_{t_2}$ (for example) and the range will be incorrectly estimated. Consequently, the 'maximum unambiguous range' is determined by the 'pulse repetition frequency' $f_p$ as

$$R_{max} = \frac{c + v}{2 f_p}. \qquad (2.5\text{--}2)$$

### 2.5.2  Velocity Estimation

If a target is moving with radial velocity $v$, then it will 'see' the radar wave of equation (2.2–3) in its local frame (i.e. one moving with respect to the radar). The (Galilean) coordinate transformation from the radar frame to the target frame is $R \to R' + vt$ and so the signal in the target frame will transform as

$$w(kR - \omega t) \to w(kR' + kvt - \omega t) = w(kR' - \omega' t) \qquad (2.5\text{--}3)$$

where $\omega' = \omega - kv$.

This means that the target will see a signal shifted in frequency by $-kv$ (the so-called Doppler shift) and will re-transmit (scatter) an echo waveform with frequency $\omega'$. In the frame of the target the radar will be moving with radial velocity $v$ and so, by the same line of reasoning, the radar will detect a (round-trip) echo signal shifted in frequency by $-2kv$. When a *single* frequency is transmitted, the frequency shift can be measured and the radial velocity of the target determined by dividing this frequency shift by $2k$.

### 2.5.3 The Matched Filter

Simultaneous accuracy in range and velocity estimates presents something of a problem. On the one hand, pulsewidth will determine the precision to which range estimation can be performed: the shorter the pulse, the greater the resolution. But narrow pulsewidth implies large bandwidth and this can conflict with velocity estimation for all but the simplest targets. This 'dichotomy' is expressed by the radar ambiguity function of the next section.

Real-world radar measurements will be contaminated by system noise and other non-target related signal contaminants. In the problems of pulse time-of-arrival and frequency estimation, it is very important that estimates be made as accurately as possible. The conventional systems-engineering approach to accurate measurements under these conditions is to apply a 'matched filter' which maximizes the signal-to-noise ratio at the receiver (in the presence of stationary Gaussian noise [8]) and preserves all of the information in the radar return signal [9]. To see how this works, consider the cross-correlation between a signal $s_{scatt}(t)$ measured by the radar and a 'test' signal $u(t)$:

$$\eta(t) = \int_{\mathbb{R}} s_{scatt}(t')u^*(t'+t)\,dt'. \tag{2.5-4}$$

When these signals are each scaled to unit energy, then the cross-correlation function will attain a maximum when $u = s_{scatt}$ and $t = 0$.

The signal measured by the radar will be due to the radar-transmitted field scattered from a target. This scattered field will be examined in detail in chapters 3, 6 and 7. For the time being, however, it is sufficient to model the echo signal as a time-delayed and (Doppler) frequency shifted replica of the transmission signal that has been scaled by a complex-valued amplitude $A$ which depends upon the (point) target. The matched filter therefore sets the test signal to equal the transmission signal: $u(t) = s(t)$. Inserting this into equation (2.5–4) yields the output of the matched filter as a function of the delay parameter $t$

$$\eta(t) = \int_{\mathbb{R}} s_{scatt}(t')s^*(t'+t)\,dt'. \tag{2.5-5}$$

For many practical radar systems, the output of the matched filter yields the data to be used for parameter estimation. When the point target model is accurate we can write

$$s_{scatt}(t) = A\,s(t - t_d)e^{i\vartheta(t-t_d)} \tag{2.5-6}$$

where $t_d$ is the time delay to the target and $\vartheta$ is its Doppler frequency shift, and these data are

$$\eta(t) = A\int_{\mathbb{R}} s(t')s^*(t'+t+t_d)e^{i\vartheta t'}\,dt'. \tag{2.5-7}$$

If an *extended target* is considered to be composed of $N$ separate and non-interacting points—each with amplitude $A_n$, delay $t_{d;n}$, and Doppler shift $\vartheta_n$—then the output of the matched filter will be described by

$$\eta(t) = \sum_{n=1}^{N} A_n \int_{\mathbb{R}} s(t')s^*(t'+t+t_{d;n})e^{i\vartheta_n t'} \, dt' . \tag{2.5-8}$$

Because of this, equation (2.5-5) is known as the 'radar mapping equation'.

## 2.6  THE AMBIGUITY FUNCTION

For an ideal point target, the radar mapping equation (2.5-5) expresses the output of the matched filter in factors of the target strength, position and the function

$$\chi(T, \vartheta) = \int_{\mathbb{R}} s(t')s^*(t'+T)e^{i\vartheta t'} \, dt' . \tag{2.6-1}$$

Evidently, the accuracy of the relationship between target position, radial velocity and radar data is determined by this function. The *ambiguity function* is defined as $\chi(T, \vartheta)\chi^*(T, \vartheta)$ and has become the principal tool used for describing the measurement accuracy and resolution associated with a particular radar signal. Moreover, the problem of radar imaging is really concerned with estimating the location and strength of the point scatterers that make up an extended target. Consequently, an understanding of $\chi$ will be helpful since we will eventually be interested in 'inverting' equation (2.5-8). (Formally, this inversion requires the determination of a related operator $\chi^{-1}$ which we will defer until section 8.4.)

If $s(t)$ is scaled to unit energy so that $\chi(0, 0) = 1$ then a straightforward consequence of equation (2.6-1) is

$$\left. \frac{\partial^2 |\chi(T, \vartheta)|^2}{\partial T^2} \right|_{\substack{T=0 \\ \vartheta=0}} = -2 \left[ \frac{1}{2\pi} \int_{\mathbb{R}} \omega'^2 |S(\omega')|^2 \, d\omega' \right.$$

$$\left. - \left( \frac{1}{2\pi} \int_{\mathbb{R}} \omega' |S(\omega')|^2 \, d\omega' \right)^2 \right] = -2\beta^2 \tag{2.6-2}$$

where $\beta$ is the bandwidth of the signal and the $s(t)$ in equation (2.6-1) has been Fourier transformed using equation (2.1-1). Similarly, we can define the pulsewidth $\tau$ of the signal by

$$\left. \frac{\partial^2 |\chi(T, \vartheta)|^2}{\partial \vartheta^2} \right|_{\substack{T=0 \\ \vartheta=0}} = -2 \left[ \int_{\mathbb{R}} t'^2 |s(t')|^2 \, dt' \right.$$

$$\left. - \left( \int_{\mathbb{R}} t' |s(t')|^2 \, dt' \right)^2 \right] = -2\tau^2 . \tag{2.6-3}$$

Assuming that $t|s(t)|^2$ vanishes at $t = \pm\infty$, then it is also easy to show that the cross-derivatives simplify to

$$\frac{\partial^2 |\chi(T, \vartheta)|^2}{\partial T \partial \vartheta}\bigg|_{\substack{T=0 \\ \vartheta=0}} = -2\,\mathrm{Im}\left(\int_{\mathbb{R}} t' s(t') \frac{ds^*(t')}{dt'}\,dt'\right) \tag{2.6-4}$$

$$\equiv -2\alpha\tau\beta\,.$$

The parameter $\alpha$, defined by equation (2.6-4), is known as the *error coupling coefficient* (for reasons that will shortly become apparent).

Integration by parts of the integral in equation (2.6-4) yields

$$\int_{\mathbb{R}} t' s(t') \frac{ds^*(t')}{dt'}\,dt' = t'|s(t')|^2\Big|_{-\infty}^{\infty}$$

$$- \int_{\mathbb{R}} |s(t')|^2\,dt' - \int_{\mathbb{R}} t' s^*(t') \frac{ds(t')}{dt'}\,dt'\,. \tag{2.6-5}$$

Since $s(t)$ is scaled to unit energy, we can conclude

$$\mathrm{Re}\int_{\mathbb{R}} t' s(t') \frac{ds^*(t')}{dt'}\,dt' = -\tfrac{1}{2}\,. \tag{2.6-6}$$

Comparing this result with that of equation (2.6-4) yields

$$\left|\int_{\mathbb{R}} t' s(t') \frac{ds^*(t')}{dt'}\,dt'\right|^2 = \tfrac{1}{4} + \alpha^2\tau^2\beta^2\,. \tag{2.6-7}$$

Applying the Schwartz inequality to the left-hand side of equation (2.6-7), and employing equations (2.1-6) and (2.6-3), results in the *radar uncertainty principle*:

$$(\tau^2 + \bar{t}^2)(\beta^2 + \bar{\omega}^2) \geqslant \tfrac{1}{4} + \alpha^2\tau^2\beta^2 \tag{2.6-8}$$

where $\bar{t} = \int_{\mathbb{R}} t'\,|s(t')|^2\,dt'$. (Note that when $\alpha = \bar{\omega} = \bar{t} = 0$ we obtain the 'usual' form of the uncertainty relation as $\tau\beta \geqslant \tfrac{1}{2}$.)

The uncertainty relation is not nearly as useful as one obtained by examining the behavior of the ambiguity function in the neighborhood of its main peak (which occurs at the origin). In terms of equations (2.6-2)–(2.6-4), the Taylor series expansion for $|\chi(T, \vartheta)|^2$ near the origin becomes

$$|\chi(T, \vartheta)|^2 \approx |\chi(0, 0)|^2$$

$$+ \frac{\partial^2 |\chi|^2}{\partial T^2}\bigg|_{\substack{T=0 \\ \vartheta=0}} T^2 + 2\,\mathrm{Re}\,\frac{\partial^2 |\chi|^2}{\partial T \partial \vartheta}\bigg|_{\substack{T=0 \\ \vartheta=0}} T\vartheta + \frac{\partial^2 |\chi|^2}{\partial \vartheta^2}\bigg|_{\substack{T=0 \\ \vartheta=0}} \vartheta^2 \tag{2.6-9}$$

$$= 1 - 2\left(\beta^2 T^2 + 2\alpha\beta\tau T\vartheta + \tau^2\vartheta^2\right)\,.$$

The ambiguity function is primarily a tool for establishing the accuracy of radar measurements in the presence of noise. And, while detailed analysis of system noise is beyond the scope of this overview, the central idea is easy to see from this expansion. Since the output of the matched filter is given by $\chi(T, \vartheta)$, the ambiguity function $|\chi(T, \vartheta)|^2$ will be proportional to the energy of the radar echo pulse.

If $N^2$ denotes the (suitably normalized) noise energy, then $|\chi(T, \vartheta)|^2 - N^2$ will be the 'usable' energy for target parameter estimation—the larger the noise energy, the poorer the performance we can expect from any estimation algorithm. Comparing this result with equation (2.6–9) allows us to identify an estimation threshold by

$$\beta^2 T^2 + 2\alpha\beta\tau T\vartheta + \tau^2\vartheta^2 = \frac{N^2}{2}. \qquad (2.6\text{--}10)$$

This is the equation for an ellipse and is used to describe the effect of measurement noise on radar performance. For this reason, equation (2.6–10) is known as the 'uncertainty ellipse'.

It can be shown that when the noise is Gaussian, the error variance in the estimation of the delay $T$ is

$$\langle (T - \overline{T})^2 \rangle = \frac{2N^2}{\beta^2} \qquad (2.6\text{--}11)$$

where $\langle \cdot \rangle$ denotes probabilistic 'expectation'. Similarly

$$\langle (\vartheta - \overline{\vartheta})^2 \rangle = \frac{2N^2}{\tau^2}. \qquad (2.6\text{--}12)$$

Equations (2.6–11) and (2.6–12) represent the minimum error variance and are twice the values of the semi-axes of the uncertainty ellipse. The uncertainty ellipse is rotated because of the cross-terms, however, and so the *maximum values* of the measurement uncertainties will be given by (see [1] for details)

$$\Delta T_{max} = \frac{\sqrt{2}N}{\beta} \frac{1}{\sqrt{1 - \alpha^2}} \qquad (2.6\text{--}13)$$

and

$$\Delta\vartheta_{max} = \frac{\sqrt{2}N}{\tau} \frac{1}{\sqrt{1 - \alpha^2}}. \qquad (2.6\text{--}14)$$

(This is illustrated in figure 2.3.)

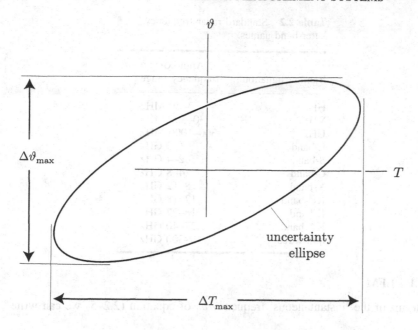

**Figure 2.3** Relationship between the uncertainty ellipse and the measurement error variance for Gaussian noise.

## 2.7  RADAR MEASUREMENT SYSTEMS

Radar systems are often broadly classified in terms of the frequency band of the waveforms that they transmit. These frequency bands are commonly distinguished by letter designations which were originally devised for military secrecy but have now come into popular usage (despite their somewhat obscure nature—see table 2.2). In addition, because of the relationship between the ambiguity function and parameter estimation in the presence of noise, a specific modulation technique is usually chosen to solve some detection problem (within the constraints of the system design). For this reason, radar systems are often further classified by the form of signal modulation that they use to create their transmitted pulses. It is the 'modulation-type' classification that is the more useful to us here, and many distinctive waveform variations have been developed since the first continuous-wave (CW) measurements were made in 1922. (Many of the resultant ambiguity functions are discussed and cataloged in [1, 2].) Rather than attempt a comprehensive survey of modern radar system types, we will only discuss two illustrative methods—one practical and one ideal.

**Table 2.2**  Standard radar-frequency letter-band names.

| Band designation | Approximate frequency range |
|---|---|
| HF | 3–30 MHz |
| VHF | 30–300 MHz |
| UHF | 300–1000 MHz |
| L-band | 1–2 GHz |
| S-band | 2–4 GHz |
| C-band | 4–8 GHz |
| X-band | 8–12 GHz |
| KU-band | 12–18 GHz |
| K-band | 18–27 GHz |
| KA-band | 27–40 GHz |
| mm-wave | 40–300 GHz |

### 2.7.1  LFM

In terms of the 'instantaneous' frequency $\omega_i$ of equation (2.2–5) we can write

$$\Phi(t) = \int_0^t \omega_i(t') \, dt'. \tag{2.7-1}$$

Linear Frequency Modulation (LFM) systems alter the instantaneous frequency in a linear fashion so that $\omega_i = \omega_0 + \gamma t$. For this kind of phase modulation we have

$$\Phi(t) = \omega_0 t + \tfrac{1}{2}\gamma t^2. \tag{2.7-2}$$

When $a(t) = \mathrm{rect}(t)$, LFM systems are also called 'chirp' radar systems. The matched filter response to a chirp signal with pulsewidth $\tau$ is given by

$$\chi(T, \vartheta) = e^{-i(2\omega_0 + \vartheta + \gamma\tau)T/2} \\ \times \tau^{-1}(\tau - |T|)\,\mathrm{sinc}\left[\tfrac{1}{2}(\tau - |T|)(\vartheta - \gamma T)\right], \quad |T| \leqslant \tau \tag{2.7-3}$$

where $\mathrm{sinc}(u) = u^{-1}\sin u$. From this response, we can determine the ambiguity function and describe the behavior of an LFM system in problems of range and range-rate estimation. In particular, we can see from equation (2.6–2) that the bandwidth is found to be

$$\beta^2 = -\frac{1}{2}\left.\frac{\partial^2 |\chi(T, \vartheta)|^2}{\partial T^2}\right|_{\substack{T=0 \\ \vartheta=0}} = \frac{1}{\tau^2} + \frac{(\tau\gamma)^2}{12}. \tag{2.7-4}$$

A more interesting and immediate issue concerns the relationship between the instantaneous frequency $\omega_i$ and the 'Fourier frequency' $\omega$ defined by equation

(2.1–1). The spectrum of a chirp signal of pulsewidth $\tau$ is given by $|S(\omega)|^2$ where

$$S(\omega) = \frac{1}{\sqrt{\tau}} \int_{-\tau/2}^{\tau/2} \exp\left\{i\left[\omega_0 t' + \tfrac{1}{2}\gamma\left(t' + \tau/2\right)^2\right]\right\}e^{-i\omega t'}\,dt'. \qquad (2.7\text{–}5)$$

This integral can be expressed in terms of Fresnel integrals $C(x)$ and $S(x)$. Performing the integration yields

$$|S(\omega)|^2 = \frac{\pi}{\tau\gamma}\,|(C(\xi) - C(\zeta)) + i\,(S(\xi) - S(\zeta))|^2 \qquad (2.7\text{–}6)$$

where

$$\xi = \tau\sqrt{\frac{\gamma}{2}} + \frac{\omega_0 - \omega}{\sqrt{2\gamma}} \qquad \text{and} \qquad \zeta = \frac{\omega_0 - \omega}{\sqrt{2\gamma}}. \qquad (2.7\text{–}7)$$

Figure 2.4 is a plot of $|S(\omega)|^2$ for a chirp with $\tau = 5$ and $\gamma = 50$.

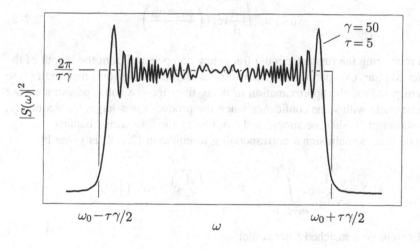

**Figure 2.4**   Spectrum for a chirp pulse of length $\tau$.

In practice, the 'spectral content' of the chirp signal is not usually determined by calculating the Fourier transform as in equation (2.7–5). Rather, since the instantaneous frequency is a linear function of time, the 'spectral' strength is determined by sampling the signal $s(t)$ in time over the pulsewidth—successive time-samples corresponding to successive (instantaneous) frequencies. The Fourier components of equation (2.7–5) are defined over an infinite time domain while the 'instantaneous' frequencies of equation (2.2–5) are defined as the time derivative of the phase. Measurement systems that estimate frequency by heterodyning one signal against another (so that the phases themselves are compared as functions of time) are actually measuring the instantaneous frequency of equation (2.2–5). In general, $\omega$ and $\omega_i$ are not the same quantities. They are related, however, and it is known that the first moment of the Fourier frequency corresponds to the first moment of the instantaneous frequency. Moreover, when $a(t)$ varies much more slowly than $\phi(t)$, the second moments also coincide [10].

### 2.7.2   An 'Idealized' $\chi$ for Radar Imaging

It is notationally inconvenient to explicitly write the radar data in terms of the ambiguity function associated with a specific signal. Also, general radar image processing algorithms can usually be modified to account for the fact that measured data will depend on some $\chi(T, \vartheta)$. For this reason, it has become common for image processing schemes to suppress the factor of $\chi(T, \vartheta)$ by (effectively) selecting an idealized radar signal of the form

$$S(\omega_i) = \sqrt{\frac{2\pi}{\beta}} \operatorname{rect}\left(\frac{\omega_i - \omega_0}{\beta}\right) \qquad (2.7\text{–}8)$$

and measuring the (instantaneous) frequency components from the output of the matched filter. Comparing this idealization with figure 2.4 shows that, in the case of chirp radars, the approximation $\omega = \omega_i$ may require some adjustment, but can be made with some confidence when the product $\tau\gamma$ is large. Consequently, the subscript 'i' shall be suppressed from $\omega_i$ in the following chapters.

The time-domain signal corresponding to equation (2.7–8) is given by

$$s(t) = \frac{1}{\sqrt{2\pi\beta}} \int_{\omega_0 - \beta/2}^{\omega_0 + \beta/2} e^{i\omega' t} \, d\omega' = \sqrt{\frac{\beta}{2\pi}} e^{i\omega_0 t} \operatorname{sinc}\left(\tfrac{1}{2}\beta t\right) \qquad (2.7\text{–}9)$$

and results in a matched filter response

$$\begin{aligned} \chi(T, \vartheta) &= e^{-i(\omega_0 + \vartheta/2)T} \\ &\times \beta^{-1} (\beta - |\vartheta|) \operatorname{sinc}\left[\tfrac{1}{2}(\beta - |\vartheta|)T\right], \quad |\vartheta| \leqslant \beta. \end{aligned} \qquad (2.7\text{–}10)$$

This (equivalent) time-domain signal cannot actually be realized by an actual radar (since it has infinite support), but the practical difficulties associated with observation are usually inconsequential.

### 2.7.3   Monopulse Tracking

In addition to estimating range and velocity (range-rate), radar systems can also be used to determine target bearing (target angular position with respect to the radar). When good accuracy is required, target bearing is usually measured by so-called 'monopulse' radar systems [12,13]. The basic physical principle is simple to understand: the waveform scattered from a simple *point* target will have constant phase surfaces ('phase fronts') which consist of concentric spherical shells centered on the target and the (inward pointing) normal to these phase fronts will be directed at the target. Monopulse systems are designed to estimate these normals.

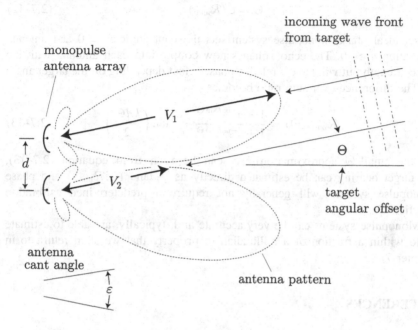

**Figure 2.5** Monopulse tracking radar.

There are two basic types of monopulse radar measurement methods—phase comparison and amplitude comparison—and either idea can be explained by considering the general configuration of figure 2.5 which illustrates an idealized one-dimensional tracking system.

Denote the (complex) voltages measured in each antenna by $V_i e^{i\phi_i}$, $i = 1, 2$. In an amplitude monopulse system, the antenna separation $d = 0$ (they have the same 'phase centers') and the antennas are canted with respect to a common axis by some non-zero angle $\varepsilon$. In this case, the voltage that each antenna measures from a plane-wave echo pulse will have different amplitudes and identical phases. This is because the plane wave will couple to each antenna by amounts determined by respective antenna patterns (which are also canted). The difference in voltage measured by each antenna will depend on the target's position offset angle $\Theta$. A monopulse system forms the 'difference-over-sum' ratio as its fundamental measurable. For the amplitude monopulse case, this ratio becomes

$$\mathcal{R}_{\text{amp}}(\Theta) = \frac{V_2 - V_1}{V_2 + V_1}. \tag{2.7-11}$$

These measurements are then (generally) compared with a calibration function $C(\mathcal{R})$ to obtain the target bearing as

$$\Theta = C(\mathcal{R}_{\text{amp}}). \tag{2.7-12}$$

An ideal phase monopulse system sets the cant angle $\varepsilon = 0$ and antenna separation $d > 0$. The echo voltages now couple into each antenna with the same strength but will have a relative phase offset depending on the target angle $\Theta$. The difference-over-sum ratio becomes

$$\mathcal{R}_{\text{phase}}(\Theta) = \frac{1 - e^{-ikd \sin \Theta}}{1 + e^{-ikd \sin \Theta}} \approx i \tan\left(\frac{kd\Theta}{2}\right) \tag{2.7-13}$$

(in the small-$\Theta$ approximation). As can be seen from equation (2.7–13), the target bearing can be estimated directly as $\frac{2}{kd} \tan^{-1}(-i\mathcal{R}_{\text{phase}})$ and phase monopulse systems will generally not require a predetermined calibration function $C$.

Monopulse systems can be very accurate and, typically, are able to estimate $\Theta$ to within a fraction of a milliradian (a property that we shall return to in chapter 7).

# REFERENCES

[1]  Cook C E and Bernfeld M 1967 *Radar Signals* (New York: Academic)
[2]  Rihaczek A W 1969 *Principles of High-Resolution Radar* (New York: McGraw-Hill)
[3]  Skolnik M L 1980 *Introduction to Radar Systems* 2nd edn (New York: McGraw-Hill)
[4]  Levanon N 1988 *Radar Principles* (New York: Wiley)
[5]  Berkowitz R S *Modern Radar* (New York: Wiley)
[6]  Guenther B D, Bennett J J, Gamble W L, and Hartman R L 1976 Submillimeter research: a propagation bibliography *US Army Missile Command, Redstone Arsenal, Alabama, Technical Report* RR-77-3 (distribution unlimited)
[7]  Nuttall A H 1966 On the quadrature approximation to the Hilbert transform of modulated signals *IEEE Proc.* **54** 1458
[8]  North D O 1943 An analysis of the factors which determine signal-noise discrimination in pulsed carrier systems *RCA Lab. Report* PTR-6C
[9]  Woodward P M 1953 *Probability and Information Theory* (New York: Pergamon)
[10] Mandel L 1974 Interpretation of instantaneous frequency *Am. J. Phys.* **42** 840
[11] Burdic W S 1968 *Radar Signal Analysis* (Englewood Cliffs, NJ: Prentice-Hall)
[12] Rhodes D R 1959 *Introduction to Monopulse* (New York: McGraw-Hill)
[13] Sherman S M 1984 *Monopulse Principles and Techniques* (Dedham, MA: Artech House)

# 3

---

# Scattering Models

In the last chapter we overviewed the measurement end of the radar echo-location scheme. We saw that the ability of a radar to estimate target range, speed and bearing is a consequence of the use of waveforms which satisfy the wave equation and, while we did not explicitly say so, these waveforms are also magnetic field $H$ or electric field $E$ solutions to the Maxwell equations. Of course, the problem of radar-based target *imaging* goes beyond ordinary target echo-location, and an understanding of image-formation algorithms will also require an examination of the scattering end of the radar process.

Traditionally, scattering models are formed from the reduced wave equation (the Helmholtz equation) for the (frequency domain) field $H(x)$ (or $E(x)$):

$$\nabla^2 H(x) + k^2 (1 + \rho(x)) H(x) = 0 \,.$$

In this equation, $k$ is a positive number and $\rho(x)$ is smooth with support $D$ (these quantities will be defined below). In the scattering problem the field $H$ can be written in the form $H = H_{\text{inc}} + H_{\text{scatt}}$ and solutions are sought which obey

$$\nabla^2 H_{\text{inc}}(x) + k^2 H_{\text{inc}}(x) = 0$$

$$\lim_{r \to \infty} \left( \frac{\partial H_{\text{scatt}}}{\partial r} - ik H_{\text{scatt}} \right) = 0$$

(the free-space Helmholtz equation and the Sommerfeld radiation condition, respectively).

The direct scattering problem seeks to determine $H_{\text{scatt}}$ when $\rho$ and $H_{\text{inc}}$ are pre-specified. The solution to the direct problem can be written in the form of a Lippmann–Schwinger equation

$$H_{\text{scatt}}(x) = k^2 \int_D G_k(x, x') \rho(x') \left( H_{\text{inc}}(x') + H_{\text{scatt}}(x') \right) \, dx'$$

where $G_k(x, x')$ is the free-space Green's function for the Helmholtz equation.

The inverse problem—determination of $\rho$ given $H_{\text{inc}}$ and $H_{\text{scatt}}$—can be attacked by using the Lippmann–Schwinger equation as a model for $H_{\text{scatt}}$ which is considered to be 'parametrized by $\rho$'. For example, if we expand $\rho$ in terms of the basis $\{\phi_j\}$ so that $\rho = \sum_{j=1}^{N} a_j \phi_j$, then we obtain the model $H_{\text{scatt}}(\boldsymbol{x}; a_1, \ldots, a_N)$ with parameters $\{a_j\}$ which can be fit to the measured data. This approach is not completely satisfying, however, since efficient modeling of the function $\rho$ often requires a good knowledge of its properties (i.e. which basis $\{\phi_j\}$ is 'efficient'?). In addition, these equations are nonlinear ($H_{\text{scatt}}$ is determined by an integral of itself) and computationally intensive. As we shall see, it is sometimes possible to linearize problems of this kind by dropping the $H_{\text{scatt}}(\boldsymbol{x}')$ term that occurs under the integral sign. This 'weak scatterer' approximation greatly reduces the computational burden required for calculating $H_{\text{scatt}}$ but, occasionally, results in model-induced artifacts in the resulting estimate of $\rho$ (the 'image'). In order that these artifacts be understood and, thereby, possibly mitigated, we will need a more complete understanding of $\rho$ and how it is related to the radar target in question.

By virtue of the fact that the interrogating and echo waveforms will satisfy the Maxwell equations, an ideal radar system will actually transmit a *vector-valued* incident pulse $\boldsymbol{H}_{\text{inc}}(\boldsymbol{x}, t)$ and measure (components of) a vector-valued echo $\boldsymbol{H}_{\text{scatt}}(\boldsymbol{x}, t)$ reflected from the target. This complication can be formally circumvented by analysing the problem component-wise but, since these components will generally be coupled, we prefer a vector approach throughout (although the true justification for this will not come until section 8.2). In addition we note that the transmitter and receiver can generally lie at different locations in space and the 'bistatic angle' is defined as the angle between these two radar parts as seen from the target's perspective. This additional parameter is not necessary in our discussion, however, and serves only to complicate some of our results. Moreover, the central radar imaging issues can be developed using the 'monostatic' assumption in which the transmitter and receiver are co-located (zero bistatic angle) and this is the situation that we shall consider below.

## 3.1    THE MAGNETIC FIELD INTEGRAL EQUATION FOR A PERFECT CONDUCTOR

We begin by developing an integral equation suitable for the determination of $H_{\text{scatt}}$ when the target is a perfect conductor $D \subset \mathbb{R}^3$ fixed in free space. Physically, an electromagnetic wave incident upon $D$ will excite currents $\boldsymbol{J}(\boldsymbol{x}, t)$ at time $t$ and position $\boldsymbol{x}$ on the boundary $\partial D$ of $D$. These currents will create a response field—the *scattered field*—which is to be measured and used to determine $\boldsymbol{J}$ (see figure 3.1). Information about $\partial D$ is encoded into $H_{\text{scatt}}$ through $\boldsymbol{J}$. Usually, of course, $\boldsymbol{J}$ will not be restricted to $\partial D$ and we must rely on some additional model if we are to estimate $\partial D$. Here, the perfect

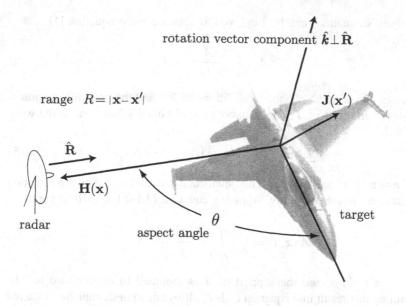

rotation vector component $\hat{k} \perp \hat{R}$

range $R = |\mathbf{x} - \mathbf{x}'|$

$J(\mathbf{x}')$

$\hat{R}$

$H(\mathbf{x})$

radar

$\theta$

aspect angle

target

**Figure 3.1** Radar data collection geometry.

conductor assumption (no currents within $D$) is used because it is both tractable and applicable to most common radar targets.

The boundary conditions for a perfect conductor are based on the assumption that the charges inside the conductor are able to respond instantaneously to changes in the fields and produce the response current $J$ and surface-charge density $\sigma$. Let $\hat{n}$ denote the unit normal to $\partial D$ at $x_b$. The perfect conductor boundary conditions for the magnetic field $H$ and the electric field $E$ require that only the normal components of $E$ and the tangential components of $H$ can exist near $\partial D$. Specifically, we have

$$\hat{n} \times H(x_b, t) = J(x_b, t)$$
$$\hat{n} \times E(x_b, t) = 0$$
$$\hat{n} \cdot H(x_b, t) = 0 \tag{3.1-1}$$
$$\hat{n} \cdot E(x_b, t) = \sigma(x_b, t).$$

Define a vector potential $A$ and a scalar potential $\varphi$ by

$$H = \nabla \times A \qquad \text{and} \qquad E = -\nabla\varphi - \frac{1}{c}\frac{\partial A}{\partial t} \tag{3.1-2}$$

where $c$ is the electromagnetic propagation speed and $A$ and $\varphi$ are related through (the Lorentz gauge)

$$\nabla \cdot A + \frac{1}{c}\frac{\partial \varphi}{\partial t} = 0. \tag{3.1-3}$$

Maxwell's equations can be employed to obtain a wave equation [1]

$$\nabla^2 A - \frac{1}{c^2}\frac{\partial^2 A}{\partial^2 t} = -J \qquad (3.1\text{--}4)$$

which, together with the boundary conditions on $\partial D$ and the radiation condition, expresses $A$ in terms of $J$. The 'outgoing wave' Green's function for the wave equation is

$$g(x, t; x', t') = \frac{1}{4\pi r}\delta\left[t' - (t - r/c)\right] \qquad (3.1\text{--}5)$$

where $r \equiv x - x'$ and $r = |r|$. This 'elemental solution' satisfies the radiation condition and assures causality. Applying equation (3.1–5) to (3.1–4) yields

$$A(x, t) = \frac{1}{4\pi}\int_{\partial D}\frac{J(x', t')}{r}\,\mathrm{d}S' \qquad (3.1\text{--}6)$$

where $t' \equiv t - r/c$ and the support of $J$ is assumed to be confined to $\partial D$. Substituting this result into equation (3.1–2) allows us to determine the scattered field from the induced currents:

$$H_{\text{scatt}}(x, t) = \nabla \times A(x, t) = \frac{1}{4\pi}\int_{\partial D}\mathcal{L}_r\{J\}(x', t') \times \hat{r}\,\mathrm{d}S' \qquad (3.1\text{--}7)$$

where $\mathcal{L}_r\{J\}(x', t') \equiv \left(r^{-2} + (rc)^{-1}\partial/\partial t'\right)J(x', t')$.

The relevant boundary condition is $J = \hat{n} \times H_{\text{tot}} \equiv \hat{n} \times (H_{\text{inc}} + H_{\text{scatt}})$. If we assume that the surface $\partial D$ is smooth (so that $\hat{n}$ is well-defined), then when $x \in \partial D$ we have

$$J(x, t) = \hat{n} \times H_{\text{inc}}(x, t) + \hat{n} \times \frac{1}{4\pi}\int_{\partial D}\mathcal{L}_r\{J\}(x', t') \times \hat{r}\,\mathrm{d}S'. \qquad (3.1\text{--}8)$$

The integral equation (3.1–8) has a weak singularity that must be accounted for when $x' \to x$. The integral can be evaluated by deforming the surface $\partial D$ in the neighborhood of $x$ and considering the limit as the deformity vanishes. We apply the hemispherical deformity of radius $a$ illustrated in figure 3.2. Expanding the triple vector product in equation (3.1–8) allows us to write

$$\hat{n} \times \frac{1}{4\pi}\int_{\partial D}\mathcal{L}_r\{J\}(x', t') \times \hat{r}\,\mathrm{d}S' \qquad (3.1\text{--}9)$$

$$= \lim_{a \to 0}\frac{1}{4\pi}\left(\hat{n} \times \int_{\partial D\backslash\sigma_a}\mathcal{L}_r\{J\}(x', t') \times \hat{r}\,\mathrm{d}S' + \int_{\sigma_a}(\hat{n}\cdot\hat{r})\mathcal{L}_r\{J\}(x', t')\,\mathrm{d}S'\right)$$

where $\sigma_a$ denotes the surface of the hemisphere of radius $a$, and we have used the fact that $\hat{n} \cdot J(x, t) = 0$.

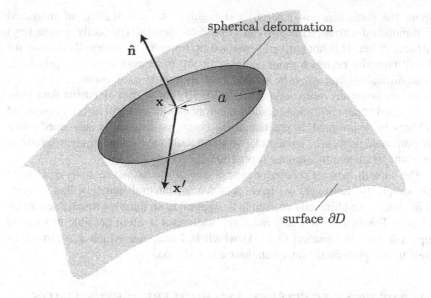

**Figure 3.2** Deformed surface geometry for the evaluation of the integral in equation (3.1–9).

The smooth-surface assumption enables us to conclude that

$$\lim_{a \to 0} \int_{\sigma_a} \frac{\hat{n} \cdot \hat{r}}{r} \, dS' = \lim_{a \to 0} a \int_{\sigma_a} d\Omega = 0$$

and                                                                                    (3.1–10)

$$\lim_{a \to 0} \int_{\sigma_a} \frac{\hat{n} \cdot \hat{r}}{r^2} \, dS' = \lim_{a \to 0} \int_{\sigma_a} d\Omega = 2\pi \, .$$

($d\Omega$ denotes the incremental solid angle subtended by $dS$.) Application of these results to (3.1–9) shows that the deformation term (the integral over $\sigma_a$) is equal to $\frac{1}{2} J(x, t)$. Subtracting this term from both sides of equation (3.1–8) yields a space-time integral equation for $J$ when this current is restricted to $\partial D$:

$$J(x, t) = 2\hat{n} \times H_{\text{inc}}(x, t) + \hat{n} \times \frac{1}{2\pi} \fint_{\partial D} \mathcal{L}_r \{J\} (x', t') \times \hat{r} \, dS' \quad (3.1–11)$$

where $\fint(\cdots)$ denotes the principal value of the integral. In principle, equation (3.1–11) can be solved for $J$ and equation (3.1–7) can be used to obtain the scattered field.

When $\partial D$ is not smooth then equation (3.1–11) must be modified. This will alter the strength (but not the form) of the current $J$ at points $x$

where the surface is not smooth, and highly localized regions of increased or diminished current strength can appear on general (perfectly conducting) surfaces. When $D$ is not perfectly conducting, however, the integral equation for $J$ will typically be much more complex [2, 3]. For direct scattering problems, these difficulties can often be treated in a straightforward manner. In inverse problems, however, these complications usually mean that extensive data must be obtained under controlled conditions before any meaningful information about $\partial D$ may be determined. In practice, these issues are frequently subsumed into a phenomenological approach to radar scattering and have been largely responsible for many extensive measurement programs.

The smooth, perfect conductor case is one of the easiest to analyse. Despite this simplicity, however, we have obtained an integral equation that is quite difficult to solve efficiently enough to be applied to an iterative inverse scattering solution. Fortunately, in many radar applications it is often possible to form an approximation to equation (3.1–11) which is linear and which accommodates itself to straightforward interpretations and solution.

## 3.2  THE WEAK SCATTERER AND HIGH-FREQUENCY LIMITS

The integral term in equation (3.1–11) accounts for multiple scattering events. When the total contribution to $J$ due to these subsequent field interactions is small in comparison with the initial driving field $2\hat{n} \times H_{\text{inc}}$ we can approximate

$$J(x, t) \approx J_{\text{po}}(x, t) = 2\hat{n} \times H_{\text{inc}}(x, t). \qquad (3.2–1)$$

This is the so-called 'weak scatterer' approximation. The 'physical optics' field scattered by $D$ is defined using equation (3.1–7) as an integral over the illuminated portion of $\partial D$ by

$$
\begin{aligned}
H_{\text{po}}(x, t) &\equiv \frac{1}{4\pi} \int_{\partial D} \mathcal{L}_R \left\{ J_{\text{po}} \right\} (x', t') \times \hat{R} \, \mathrm{d}S' \\
&= \frac{1}{2\pi Rc} \int_{\hat{R}\cdot\hat{n}<0} \hat{R} \cdot \hat{n} \frac{\partial H_{\text{inc}}(x', t')}{\partial t'} \, \mathrm{d}S' + \mathrm{O}\left(R^{-2}\right),
\end{aligned}
\qquad (3.2–2)
$$

where $R = x - x'$ and we have assumed the target dimensions are small with respect to $R$ so that the $R^{-1}$ factor can be brought out from under the integral sign. The last relationship in equation (3.2–2) has made use of the triple vector product identity and the fact that $\hat{R} \cdot H_{\text{inc}} = 0$. The physical optics *far field* approximation, appropriate to radar problems, drops the $\mathrm{O}\left(R^{-2}\right)$ terms.

The frequency-domain representation of equation (3.2–2) can be obtained by simply considering a plane wave incident interrogating waveform $H_{\text{inc}}(x, t) = H_0 \exp[\mathrm{i}(k\hat{R} \cdot x + kR - \omega t)]$ with frequency $\omega = ck$. Substitution yields

$$H_{\text{po}}(x, t; k) = -\frac{\mathrm{i}kH_0 \mathrm{e}^{\mathrm{i}(2kR-\omega t)}}{2\pi R} \int_{\hat{R}\cdot\hat{n}<0} \hat{R} \cdot \hat{n} \mathrm{e}^{\mathrm{i}2k\hat{R}\cdot x'} \, \mathrm{d}S'. \qquad (3.2–3)$$

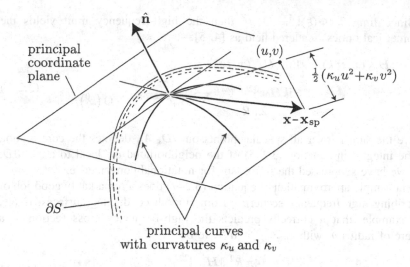

**Figure 3.3**  Principal coordinates for use in evaluating equation (3.2–5).

(The factor of 2 accounts for the two-way travel from radar to target and back.)

When $\partial D$ is smooth and $k$ is large the integral in equation (3.2–3) can be evaluated by the method of stationary phase. This method relies on rapid phase variations to make the integrand average to almost zero when integrated over all points except those around which the phase takes on an extremal value. In the case of equation (3.2–3), the extremal values will be those for which $\hat{R} \cdot x' = 0$ or, equivalently, those points on $\partial D$ at which $\hat{n} \parallel \hat{R}$. Expanding $\partial D$ about one of these so-called *specular points* $x_{sp}$ in local principal coordinates $u$ and $v$ yields

$$\hat{R} \cdot x \approx \hat{R} \cdot x_{sp} + \tfrac{1}{2}\left(\kappa_u u^2 + \kappa_v v^2\right) \tag{3.2–4}$$

where $\kappa_u$ and $\kappa_v$ are the principal curvatures of the surface at $x_{sp}$ (see figure 3.3). Substituting this expansion into equation (3.2–3) allows us to approximate the contribution to the scattered field from the small patch of surface $\sigma_m = \{(u, v)|(u, v) \in [-a, a]_m \times [-b, b]_m\}$ surrounding $x_{sp;m} \in \partial D$ by

$$H_{po;m}(x, t; k) \approx -\frac{ik H_0 e^{i(2kR-\omega t)}}{2\pi R}$$
$$\times e^{i2k\hat{R}\cdot x_{sp;m}} \int_{\sigma_m} \exp\left[ik\left(\kappa_u u^2 + \kappa_v v^2\right)\right] du\,dv. \tag{3.2–5}$$

This last term is expressible as a product of Fresnel integrals $\Phi(\xi) = C(\xi) + iS(\xi)$ and we can write

$$H_{po;m}(x, t; k) \approx -\frac{ik H_0 e^{i(2kR-\omega t)}}{2\pi R}$$
$$\times e^{i2k\hat{R}\cdot x_{sp;m}} \frac{2\pi}{\sqrt{\kappa_u \kappa_v}} \Phi\left(a_m\sqrt{k\kappa_u}\right) \Phi\left(b_m\sqrt{k\kappa_v}\right). \tag{3.2–6}$$

Since $\lim_{\xi \to \infty} |\Phi(\xi)| = 2^{-1/2}$ then the high-frequency limit yields the 'geometrical optics' scattered field as [4, 5]:

$$\boldsymbol{H}_{\text{go}}(\boldsymbol{x}, t; k) \equiv \lim_{\substack{\text{large } k}} \boldsymbol{H}_{\text{po}}(\boldsymbol{x}, t; k)$$

$$= -\frac{ik\boldsymbol{H}_0 e^{i(2kR-\omega t)}}{2\pi R} \sum_m A_m e^{i2k\hat{R}\cdot\boldsymbol{x}_m} + O(k^0) \qquad (3.2\text{--}7)$$

where the sum is over all specular points on $\partial D$, $A_m$ denotes the contribution to the integral in equation (3.2–5) of the neighborhood $\sigma_m$ local to $\boldsymbol{x}_m \in \partial D$, and we have suppressed the suffix 'sp' for notational convenience.

The simple approximation of equation (3.2–7) does a remarkably good job of describing high-frequency scattering from smooth conducting surfaces. (Note, for example, that it correctly predicts the high-frequency cross section of a sphere of radius $a$, with $\kappa_u = \kappa_v = a^{-1}$, to be

$$\varsigma = \frac{4\pi R^2}{k^2} \frac{\|\boldsymbol{H}_{\text{go}}\|^2}{\|\boldsymbol{H}_0\|^2} = \pi a^2 \qquad (3.2\text{--}8)$$

which is, coincidentally, also the projected size of the sphere.) When the target is not a smooth scatterer (but still a weak scatterer) the contributions due to the smooth portion of $\partial D$ can be treated by equation (3.2–7) and the non-specular scattering components can be accounted for separately. For aircraft targets, these non-specular scatterers usually consist of structural seams, joints, ducts, corner reflectors and other *localized* elements (recall equation (3.1–10) and the discussion following equation (3.1–11)). In the weak scatterer approximation, it is possible to append appropriate scatterer-dependent terms to the sum in equation (3.2–7). In radar target recognition problems, however, it can be difficult to reliably distinguish the scattering mechanisms behind these terms and so they are usually all categorized as simply 'scattering centers'.

## 3.3  DIELECTRIC SCATTERERS

Many modern radar targets are composed of non-metallic composites and do not scatter electromagnetic energy as well as if they were constructed from ordinary conductors. (This fact, of course, is often the motivation for building these radar targets this way.) The integral equations of electromagnetic scattering can be developed for more than just the perfect conductor case and, as an example, we will examine scattering from a source-free, linear, homogeneous and isotropic (and non-magnetic) dielectric. The derivation used is patterned after the one for perfect conducting scatterers.

Consider the Maxwell equation for the curl of the magnetic field:

$$\nabla \times \boldsymbol{H}(\boldsymbol{x}, t) = \frac{\epsilon}{c} \frac{\partial \boldsymbol{E}(\boldsymbol{x}, t)}{\partial t} + \boldsymbol{J}(\boldsymbol{x}, t) \qquad (3.3\text{--}1)$$

where $\epsilon$ is a media-dependent parameter (the 'permittivity'). Unlike conductors, dielectrics will not support currents and, within the dielectric scatterer ($x \in D$) we set $J(x, t) = 0$. If $\epsilon_d$ denotes the permittivity of the homogeneous dielectric and $\epsilon_0$ is the permittivity of free space, then we can write equation (3.3-1) for the field within the dielectric as

$$
\begin{aligned}
\nabla \times H(x, t) &= \frac{\epsilon_d}{c} \frac{\partial E(x, t)}{\partial t} \\
&= \frac{\epsilon_0}{c} \frac{\partial E(x, t)}{\partial t} + \frac{1}{c}(\epsilon_d - \epsilon_0) \frac{\partial E(x, t)}{\partial t} \\
&= \frac{\epsilon_0}{c} \frac{\partial E(x, t)}{\partial t} + J_a(x, t)
\end{aligned}
\tag{3.3-2}
$$

where $J_a \equiv c^{-1}(\epsilon_d - \epsilon_0)\partial E/\partial t$ is the 'apparent' current [3].

Equation (3.3-2) permits us to interpret the magnetic field as existing in the region $\mathbb{R}^3$ with permittivity $\epsilon_0$ and current density $J_a$ confined to $D$. This scattering problem is, formally, the same as that considered in section 3.1 and the vector potential of equation (3.1-6) can be modified to read

$$
A(x, t) = \frac{1}{4\pi} \int_D \frac{J_a(x', t')}{r} \, dV'
\tag{3.3-3}
$$

where the support of $J_a$ is now the entire volume $D$.

The scattered field can be determined as before (equation (3.1-7)). Since $\nabla \times (\epsilon E) = \nabla \epsilon \times E + \epsilon \nabla \times E$ (and noting that the derivative of a step function is a delta function), the discontinuity in $\epsilon$ at $x \in \partial D$ results in a surface integral. It is straightforward to show

$$
\begin{aligned}
H_{\text{scatt}}(x, t) &= \nabla \times A(x, t) \\
&= \frac{\epsilon_d - \epsilon_0}{4\pi \epsilon_0} \int_{\partial D} \frac{1}{r} \hat{n} \times \frac{\epsilon_0}{c} \frac{\partial E(x', t')}{\partial t'} \, dS' \\
&\quad + \frac{\epsilon_d - \epsilon_0}{4\pi \epsilon_0} \int_D \mathcal{L}_r \left\{ \frac{\epsilon_0}{c} \frac{\partial E}{\partial t'} \right\} (x', t') \times \hat{r} \, dV'.
\end{aligned}
\tag{3.3-4}
$$

It is, of course, possible to continue following the derivation of section 3.1, but when $\epsilon_d - \epsilon_0$ is small the weak scatterer approximation will be appropriate and we can immediately linearize the scattered field integral equation by substituting the incident field into the integrand. For an incident plane wave we obtain (in the far-field)

$$
\begin{aligned}
H_{\text{scatt}}(x, t; k) &= -\frac{ik H_0 e^{i(2kR-\omega t)}}{2\pi R} \left( \frac{\epsilon_d - \epsilon_0}{\epsilon_0} \right) \int_{\partial D} \hat{R} \cdot \hat{n} e^{i2k\hat{R}\cdot x'} \, dS' \\
&\quad + \frac{k^2 H_0 e^{i(2kR-\omega t)}}{2\pi R} \left( \frac{\epsilon_d - \epsilon_0}{\epsilon_0} \right) \int_D e^{i2k\hat{R}\cdot x'} \, dV'.
\end{aligned}
\tag{3.3-5}
$$

(This result should be compared with the perfect conductor case of equation (3.2-3).)

### 3.4  THE (APPROXIMATE) RADAR SCATTERING MODEL

The scattering center description of aircraft has proven to be very useful in practical target identification algorithms. If we define a generalized scatterer density function $\rho_{k,\hat{R}}(x)$ by

$$H_{\text{scatt}}(x, t; k) = \frac{ikH_0 e^{i(2kR-\omega t)}}{(2\pi)^3 R} \int_{\mathbb{R}^3} \rho_{k,\hat{R}}(x') e^{i2k\hat{R}\cdot x'} d^3 x' \qquad (3.4\text{--}1)$$

then we can encompass both the physical optics and geometric optics results in one weak scatterer far-field model. The scatterer density function must be interpreted differently in the physical and geometric regimes and, in general, $\rho_{k,\hat{R}}$ has a dependence on $k$ and $\hat{R}$ in either case.

$H_{\text{scatt}}$ represents the radar measurements, and if the range $R$ to the target is accurately known—or even if it is only approximately known but remains constant over the data set—then the complex-valued prefactor in this integral can be factored into these data. Since, from now on, the electromagnetic field will refer to the scattered field, we will suppress the 'scatt' subscript and write

$$H(x, t; k) = \frac{-iRe^{-i(2kR-\omega t)}}{k|H_0|} H_{\text{scatt}}(x, t; k) \qquad (3.4\text{--}2)$$

to avoid carrying the cumbersome prefactor in equation (3.4–1) throughout our discussion. We note that practicable implementation of equation (3.4–2) requires any variations in $R$ (that occur while these data are collected) to be estimable. In the one-dimensional imaging methods developed in chapter 4, $R$ will usually be constant over the data set and equation (3.4–2) rarely presents any problems. In two-dimensional imaging schemes, however, $R$ will generally vary due to the target (translational) motion that occurs during measurements made while the target rotates. The effect of failing to correctly account for this translational motion will be the introduction of phase errors into these data, and these errors can be a source of image corruption. We will return to this important problem in section 5.3.

In the sequel, we will concentrate on the principal radar methods for extracting target information from $H$ by estimating various functions closely related to $\rho_{k,\hat{R}}$. For imaging purposes it is useful to define a coordinate system fixed to the target. As the target maneuvers, this coordinate system will translate and rotate with respect to radar-centric coordinates. We assume that the target behaves as a rigid body and that its instantaneous axis of rotation has a component $\hat{k} \perp \hat{R}$. Define coordinate directions $\hat{i}$ and $\hat{j}$ in terms of the rotation angle $\theta$ by $\hat{i} \cdot \hat{R} = -\sin\theta$ and $\hat{j} \cdot \hat{R} = \cos\theta$: these are the 'cross-range' and 'down-range' directions, respectively. In terms of these coordinates, equation (3.4–1) can be written as

$$H(k, \theta) = \frac{\hat{H}_0}{(2\pi)^2} \int_{\mathbb{R}^2} \overline{\rho}_{k,\hat{R}}(x', y')$$
$$\times e^{i2k(y'\cos\theta - x'\sin\theta)} dx' dy' \qquad (3.4\text{--}3)$$

where $\boldsymbol{x}' = x'\hat{\boldsymbol{i}} + y'\hat{\boldsymbol{j}} + z'\hat{\boldsymbol{k}}$ and $\overline{\rho}_{k,\hat{R}}(x, y) \equiv 1/2\pi \int \rho_{k,\hat{R}}(x, y, z)\,\mathrm{d}z$ is the *axis-integrated* scatterer density function. Equation (3.4–3) is the 'standard' target model used in radar imaging analysis. This model is also known as the weak, far-field model.

As can be seen by inspecting equation (3.2–6), $\rho_{k,\hat{R}}$ will generally depend on the wavenumber $k$ and target orientation $\hat{R}$. For analytical convenience, however, this dependence is usually treated as insignificant over the span of $\hat{R}$ employed in, and the ranges of $k$ obtainable from, conventional radar systems and is typically ignored. This further simplification allows us to exploit equation (3.4–3) as a Fourier transform relationship between $H_{\text{scatt}}$ and $\rho = \rho_{k,\hat{R}}$. The Fourier transform relationship is very convenient for practical object function estimation but this approximation turns out to cause significant problems in image interpretation and we will return to it in chapter 6.

Another approximation that we shall (temporarily) make concerns the vector nature of the scattered field. Under the physical optics approximation of equation (3.2–1) the scattered field $H_{\text{scatt}}$ has the same polarization as the incident field $H_0$ (i.e. the radar return from a physical optics target is not depolarized). Because of this, we will suppress the vector aspect of these data until it is explicitly required (in chapter 8).

# REFERENCES

[1]  Stratton J A 1941 *Electromagnetic Theory* (New York: McGraw-Hill)
[2]  Poggio A J and Miller E K 1973 Integral equation solutions to three dimensional scattering problems *Computer Techniques for Electromagnetics* ed R Mittra (London: Pergamon)
[3]  Müller C 1969 *Foundations of the Mathematical Theory of Electromagnetic Waves* (New York: Springer)
[4]  Keller J B 1959 The inverse scattering problem in geometrical optics and the design of reflectors *IRE Trans. Antennas Propag.* **7** 146
[5]  Weiss M R 1968 Inverse scattering in the geometric-optics limit *J. Opt. Soc. Am.* **58** 1524

# 4

---

# One-Dimensional Imaging

The standard target model of equation (3.4–3) is the principal result of chapter 3 and will be used as the foundation for all of the imaging methods to be examined. Our overall approach will be to begin with the most easily implemented techniques and work up in complexity—a plan that, more or less, corresponds to the dimensionality of the resulting image. We will start with the one-dimensional case.

When it comes to the problem of articulating target substructure by radar imaging, the most straightforward methods are those that create so-called 'range profiles'. Such strategies are collectively known as high range resolution (HRR) techniques (cf [1–7] and references cited therein). HRR radar systems transmit a pulse whose instantaneous spatial support is smaller than that of the target (usually, less than about one-tenth of the target's down-range extent). As this pulse sweeps across the target it sequentially excites the target's scattering subelements which re-radiate energy back to the radar receiver. When these scattering subelements are non-interacting and point-like (i.e. can be modeled as delta functions), the scattered pulse will be a sum of damped and blurred images of the incident pulse which are shifted by time delays that are proportional to subscatterer's range (as in equation (2.5–8)).

Because of their relative simplicity—both functional and conceptual—range profile methods have a long history in radar-based target classification programs. In addition, since equation (3.4–3) can be separated into down-range and cross-range factors, we will see (in section 5.3) that range profile methods also play an important role in two-dimensional (ISAR) imaging techniques. The present discussion concentrates on some of the mathematical issues associated with stable estimates of one-dimensional target object functions from limited and noisy data.

## 4.1 RANGE PROFILES

As their name suggests, range profile images contain range-only information—they lump together, as a phasor sum, all scattering elements located at the same distance. When a single pulse is transmitted and received, a range-only image model can be formed by setting $\theta = 0$ in equation (3.4–3) and integrating over the cross-range variables $x$ and $z$. If we define

$$\bar{\bar{\rho}}(y) \equiv \frac{1}{2\pi} \int_{\mathbb{R}} \bar{\rho}(x', y)\, dx'$$

$$= \frac{1}{(2\pi)^2} \int_{\mathbb{R}^2} \rho(x', y, z')\, dx'dz' \qquad (4.1\text{–}1)$$

then equation (3.4–3) becomes

$$H(k) \equiv H(k, 0) = \frac{1}{2\pi} \int_{\mathbb{R}} \bar{\bar{\rho}}(y')e^{i2ky'}\, dy', \quad k \in [k_1, k_2]. \qquad (4.1\text{–}2)$$

Equation (4.1–2) is a frequency-domain representation. The spatial transform of this equation has a simple physical (if approximate) interpretation: a narrow pulse, traveling at speed $c$, is transmitted by the radar and reflected by the scatterers that make up $\bar{\bar{\rho}}(y) = \sum_j \bar{\bar{\rho}}_j \delta(y - y_j)$ resulting in a sequence of time delayed ($t_j = 2y_j/c$) return pulses measured by the radar receiver (see equation (4.3–3) below). The strength $\bar{\bar{\rho}}_j$ and position $y_j$ of these measured return pulses results in an image $\hat{\rho}$ which, in principle, can be used to classify/identify the target.

Figure 4.1 shows an example of the kind of one-dimensional images that can be created by HRR radar systems. The range profile is from a B-727 jetliner, a top view of which is displayed beneath (with orientation at time of measurement). These data were collected using a radar with bandwidth 500 MHz centered on 9.25 GHz ($\Delta\omega = 2\pi \times 500$ MHz and $\bar{\omega} = 2\pi \times 9.25$ GHz). An approximate mapping between some of the range profile peaks and target features is also made (vertical lines).

The simplicity of equation (4.1–2) disguises two important difficulties that must be addressed for effective radar imaging. The first problem has already been briefly hinted-at and concerns the weak, non-interacting scatterer approximation: note that the profile of figure 4.1 displays additional peaks outside of the target support ($y > 40$ m). Analysis of these 'extra' features will be deferred until chapter 6. The second problem is a consequence of the limitations in any data that will be measured by radar systems. Such data are usually corrupted by noise and are always limited by bandwidth. This second difficulty is responsible for the introduction of unwanted image artifacts and reduced image resolution, and is the subject of the remainder of this chapter.

## 4.2 ILL-POSED PROBLEMS AND REGULARIZATION

Let $\langle u, v \rangle_T = \int_T u(t)v^*(t)\, dt$ denote the inner product over $T \subset \mathbb{R}$ and define the $L^2$ norm by $\|u\|_T = \langle u, u \rangle_T^{1/2}$. If $\|u\|_T < \infty$, then $u$ is said to be square

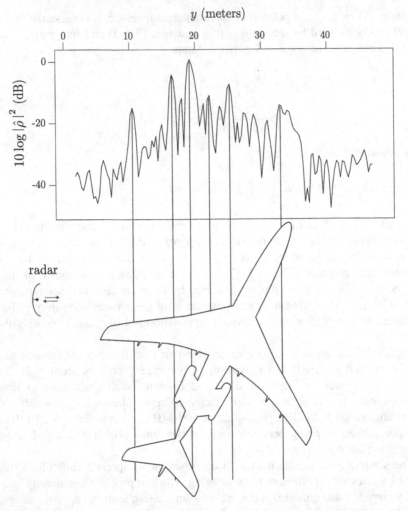

**Figure 4.1** Example range profile created using a (uniformly) weighted windowed Fourier transform of these data. The B-727 jetliner target (top view positioned below the profile) was illuminated by a radar with center frequency 9.25 GHz and bandwidth 500 MHz.

integrable over $\mathcal{T}$. The space $L^2(\mathcal{T})$ of square integrable functions over $\mathcal{T}$ is a Hilbert space and, if we consider $H \in L^2(\mathcal{K})$ and $\rho \in L^2(\mathcal{Y})$, then equation (4.1–2) can be written as

$$H = F\rho \qquad (4.2\text{–}1)$$

where $F$ is the linear operator corresponding to $H(k) = \int_{\mathcal{Y}} f(k, y')\overline{\overline{\rho}}(y')\,dy'$ which maps $\rho$ to the measurements $H$.

Equation (4.2–1) is an integral equation of the first kind. The main difficulty with first-kind equations is that they are generally ill-posed in the sense that small perturbations in $H$ can result in large and unacceptable variations in the image estimate $\overline{\overline{\rho}}_{est}$ of the object function $\overline{\overline{\rho}}$. Since any real-world measurements will contain perturbations in the form of noise, the ill-posed nature of image reconstruction can be a very troublesome practical matter [8, 9].

If we assume an additive noise model, then we can account for the corruption in the measured data by rewriting equation (4.2–1) as

$$H = F\rho + n \qquad (4.2–2)$$

where $n \in L^2(\mathcal{K})$ represents an unknown noise vector. The noise term $n$ results in additional parameters which generally make the system of equations (4.2–1) inconsistent and an approximate solution to equation (4.2–2) can be obtained by minimizing the average error:

$$\rho_{ls} = \arg \inf_{\rho \in L^2(\mathcal{Y})} \|F\rho - H\|_{\mathcal{K}}. \qquad (4.2–3)$$

These are the traditional 'least-squares' solutions and lead to the so-called 'normal equations'

$$F^{\dagger}H = F^{\dagger}F\rho_{ls} \qquad (4.2–4)$$

where $F^{\dagger}$ denotes the adjoint operator associated with $F$ (see appendix definition A.5). When $F^{\dagger}F$ is invertible (in the discrete case, when its columns are linearly independent) then the least-squares solution can be written

$$\rho_{ls} = \left(F^{\dagger}F\right)^{-1} F^{\dagger}H. \qquad (4.2–5)$$

The problem of ill-posedness is associated with the inversion of $F^{\dagger}F$ and is explained in more detail in the appendix. The results established there may be briefly stated as:

(i)   The scattering operator $F$, which maps the target scatterer density to the measured data, is a compact operator (theorem A.4).

(ii)   The least-squares estimate for the target scatterer density can be formulated in terms of the eigenfunctions and eigenvalues of the $F^{\dagger}F$ (theorems A.6 and A.10).

(iii)   These eigenvalues form a sequence that converges to zero (theorem A.7).

(iv)   The corresponding inverse of $F^{\dagger}F$ is not bounded in general (theorem A.11) and, in particular, is badly behaved when these data are finite. Increasing the dimension of the data space (by collecting more data, for example) only worsens this problem.

Not all linear integral equations are as badly behaved as equation (4.2–1) and *second-kind* integral equations turn out to be much easier to deal with than first-kind equations. Second-kind equations can be written as

$$\zeta = G\xi + \alpha\xi \qquad (4.2\text{–}6)$$

where $G$ is an operator from a Hilbert space $\mathcal{H}$ into itself and $\alpha$ is a fixed scalar. An important (utilitarian) difference between first- and second-kind equations is a consequence of the following theorem:

*Theorem* 4.2.1.    If $\alpha$ is neither 0 nor the negative of any eigenvalues of $G$, then $(G + \alpha I)^{-1}$ exists and is bounded, and equation (4.2–6) has the unique solution [10]

$$\xi = (G + \alpha I)^{-1}\zeta. \qquad (4.2\text{–}7)$$

Theorem 4.2.1 is the basis for a method that attempts to mitigate the effects of noise when solving first-kind equations. In this approach, the original problem $F^\dagger H = F^\dagger F\rho$ (equation (4.2–4)) is simply *replaced* by equation (4.2–6) by identifying $\zeta = F^\dagger H$, $\xi = \rho$ and $G = F^\dagger F$. The resulting equation

$$F^\dagger H = \left(F^\dagger F + \alpha I\right)\rho \qquad (4.2\text{–}8)$$

is a *perturbation* of equation (4.2–4) and, when $|\alpha|$ is small, the hope is that the replacement equation will be sufficiently close to the original equation that the difference between their solutions will also be small.

Methods which moderate the effects of ill-posedness by modifying the original problem are known as *regularization* techniques [11]. When a second-kind equation is used to approximate a first-kind equation, $\alpha$ is called the 'regularization parameter' and must be chosen to comply with the requirements of theorem 4.2.1. In addition, the regularization parameter must be selected to have a small enough magnitude that the solutions to the regularized equation will be representative of those of the original ill-posed equation.

An added *computational* advantage in working with second-kind integral equations is that they can often be solved without ever having to determine the inverse of $(F^\dagger F + \alpha I)$. To see this, rearrange equation (4.2–8) to read

$$\rho = \alpha^{-1}F^\dagger(H - F\rho). \qquad (4.2\text{–}9)$$

The Neumann iterative method interprets the right-hand side of this equation as an updating scheme, and an iterative procedure that expresses the approximate solution $\rho^{(j+1)}$ in terms of $\rho^{(j)}$ can be formulated as

$$\rho^{(0)} = 0$$
$$\rho^{(j+1)} = \alpha^{-1}F^\dagger(H - F\rho^{(j)}). \qquad (4.2\text{–}10)$$

This recursion reconstruction converges to the unique solution of equation (4.2–8) if $\alpha$ satisfies the condition of theorem 4.2.1 (although it may not converge very quickly).

While this kind of (straightforward) regularization method can be used to mitigate the consequences of data noise errors, it generally cannot make up for the problems associated with image reconstruction from limited data [12, 13]. Each measurement of the form of equation (4.2–1) may be considered to be the sum of projections of an unknown $\rho$ onto response vectors $f_j$. Only those components of $\rho$ that lie in the subspace spanned by the set of all $f_j$ contribute to the measurement. This subspace is called the *measurement space* and the components of $\rho$ in the subspace orthogonal to it (the null space) cannot contribute to the measurements (see appendix theorems A.8 and A.9). Any $\rho$ lying wholly in the null space will yield zero measurements.

Problems associated with the null space of $F$, which usually affect image resolution and may introduce image artifacts, result from fundamental shortfalls in the amount of 'information' that has been collected from the target in question. Resolution enhancement methods work by inserting some form of additional (unmeasured) target information into the image reconstruction process in an effort to make up for these data limitations. These techniques are often only cosmetic but, as we shall see in the next section, sometimes significant image improvement can be had by applying only minor (and reasonable) prior assumptions about the target.

Because $\|A+B\| \leqslant \|A\|+\|B\|$, the regularization method based on equation (4.2–8) results in a least-squares solution of the form

$$\rho_{\mathrm{r}} = \arg \inf_{\rho \in L^2(\mathcal{Y})} \{\|F^\dagger F\rho - F^\dagger H\|_{\mathcal{Y}} + \alpha\|\rho\|_{\mathcal{Y}}\}. \qquad (4.2\text{–}11)$$

A more general basis for regularization can be developed as an extension to solutions of equation (4.2–11). If some null-space information about the image is known, then least-squares may not provide the best overall solution. Denote this additional null-space data symbolically by $\rho_0$ and let $\mathcal{J}_2(\rho, \rho_0)$ be some functional which measures a 'difference' between $\rho$ and $\rho_0$. (For example, if $\rho_0$ is an expected image, then we could set $\mathcal{J}_2(\rho, \rho_0) = \|\rho - \rho_0\|_{\mathcal{Y}}$.) The solution given by

$$\rho_{\mathrm{r}} = \arg \inf_{\rho \in L^2(\mathcal{Y})} \{\|F^\dagger F\rho - F^\dagger H\|_{\mathcal{Y}} + \alpha\mathcal{J}_2(\rho, \rho_0)\} \qquad (4.2\text{–}12)$$

will account for the additional prior information contained in $\rho_0$. $\mathcal{J}_2$ is known as the regularization functional and $\alpha > 0$ is the regularization parameter which determines the extent to which the information in $\rho_0$ dominates that contained in $H$. By forcing the solution to lie 'close to' $\rho_0$, this approach will be a regularization technique when $\rho_0$ is properly constrained.

An obvious further modification to equation (4.2–12) extends the least-squares term to a generalized functional $\mathcal{J}_1(\rho, H)$ and seeks solutions which minimize

$$J(\rho) = J_1(\rho, H) + \alpha J_2(\rho, \rho_0). \tag{4.2–13}$$

In this form, regularization methods are the formal basis for many of the classic signal and image processing algorithms and we will return to these ideas at the end of section 4.4.

## 4.3    RESOLUTION IMPROVEMENT METHODS

Equation (4.1–2) is a Fourier transform relationship between the target object function $\overline{\overline{\rho}}$ and the data $H$:

$$H(k_y) = \mathcal{F}\{\overline{\overline{\rho}}\}(k_y) = \int_{\mathbb{R}} e^{-ik_y y'} \overline{\overline{\rho}}(y')\,dy', \qquad k_y \in \mathbb{R} \tag{4.3–1}$$

where $k_y = -2k$. Because $\overline{\overline{\rho}}(y)$ is a space-limited function (i.e. it has compact support), its Fourier transform will not be band-limited and the data set does not have finite support (this can be considered to be a consequence of the Polya–Plancherel theorem, below). Realizable measurement systems, however, can collect data only within a limited range $k \in [k_1, k_2]$ and so the inverse transform

$$\overline{\overline{\rho}}(y) = \mathcal{F}^{-1}\{H\}(y) = \frac{1}{2\pi} \int_{\mathbb{R}} e^{ik_y y} H(k_y)\,dk_y, \qquad y \in \mathbb{R} \tag{4.3–2}$$

cannot generally be used to recapture $\overline{\overline{\rho}}$ from the measured data. A simple *estimate* $\hat{\rho}(y)$ of $\overline{\overline{\rho}}(y)$ is often formed by applying windowed Fourier transform techniques, but this often proves unsatisfactory when high resolution is required. In fact, by multiplying equation (4.1–2) by $\exp(ik_y y)$ and integrating over $k_y \in [2k_1, 2k_2]$, it is easy to show that $\hat{\rho}(y)$ is related to $\overline{\overline{\rho}}(y)$ by

$$\hat{\rho}(y) = \Delta k \int_{\mathcal{Y}} \overline{\overline{\rho}}(y') \operatorname{sinc}(\Delta k(y - y'))\,dy' \tag{4.3–3}$$

where $\Delta k = k_2 - k_1$ and $\mathcal{Y} = [-L/2, L/2]$ is the support of $\overline{\overline{\rho}}$.

The function $\Delta k \operatorname{sinc}(\Delta ky) = \sin(\Delta ky)/y$ has an extended oscillatory structure and the resulting estimate $\hat{\rho}$ will be contaminated by so-called 'sidelobes' which can obscure the image of any target elements which happen to adjoin strong scatterers. The most common technique used to suppress such sidelobes is to apply a non-uniform window to these data (instead of the one implied by equation (4.1–2)) by *multiplying* $H(k)$ by some weighting function (the Hanning and Hamming functions are often enlisted for this) [14]. Non-uniform windowing can reduce sidelobes but does so at the expense of increasing the mainlobe width. Consequently, these windowing methods diminish image spatial resolution. So-called 'multiple apodization' methods, which attempt to retain the sinc mainlobe while using the sidelobes associated with other windows, have been devised and enjoy varying degrees of success.

Alternately, note that since $\lim_{\Delta k \to \infty} \Delta k \operatorname{sinc}(\Delta k y) = 2\pi \delta(y)$, the accuracy to which $\hat{\rho}$ represents $\overline{\overline{\rho}}$ will be improved (the extent of the sidelobes will be reduced) as the measurement bandwidth increases. The following theorem shows that (in principle) the bandwidth can be *effectively* increased by data extrapolation.

*Theorem* 4.3.1 *Polya–Plancherel.* The Fourier transform of a space-limited function is an analytic function that is determined completely by the values within any finite part of the Fourier domain.

## 4.3.1 Prolate Spheroidal Wave Expansion

Equation (4.3–3) is important in radar imaging because it models the field scattered from a target illuminated by an incident pulse in the spatial domain. Since the radar mapping equation (2.5–5) is of this form, equation (4.3–3) (with $t = 2y/c$) merits a more detailed examination.

Define the characteristic function of the set $S$ by

$$\mathcal{X}_S(u) = \begin{cases} 1, & \text{if } u \in S \setminus \partial S \\ \frac{1}{2}, & \text{if } u \in \partial S \\ 0, & \text{otherwise.} \end{cases} \tag{4.3-4}$$

($\mathcal{X}_S$ is a projection operator onto the interior of $S$.) Then equation (4.3–3) can be written

$$\hat{\rho}(y) = \Delta k \int_{\mathbb{R}} \mathcal{X}_y(y') \overline{\overline{\rho}}(y') \operatorname{sinc}(\Delta k(y - y')) \, dy'. \tag{4.3-5}$$

Equation (4.3–5) is a first-kind integral equation. The eigenfunctions of this equation are the prolate spheroidal wavefunctions $\psi_j$ and display the remarkable property of being orthogonal on both $\mathbb{R}$ and $\mathcal{Y}$ so that [15, 16]

$$\int_{\mathbb{R}} \psi_i(y)\psi_j(y) \, dy = \delta_{ij}$$

$$\int_{\mathbb{R}} \mathcal{X}_y(y)\psi_i(y)\psi_j(y) \, dy = \lambda_j \delta_{ij} \tag{4.3-6}$$

where $\delta_{ij}$ is the Kronecker delta. Using equation (4.3–6), we can expand $\hat{\rho}$ and $\overline{\overline{\rho}}$ as

$$\hat{\rho}(y) = \sum_{j=1}^{\infty} \langle \hat{\rho}, \psi_j \rangle \psi_j(y)$$

$$\overline{\overline{\rho}}(y) = \sum_{j=1}^{\infty} \frac{1}{\lambda_j} \langle \mathcal{X}_y \overline{\overline{\rho}}, \psi_j \rangle \mathcal{X}_y(y) \psi_j(y). \tag{4.3-7}$$

Substitution of these last two results into equation (4.3–5), and using the fact that the $\psi_j$ are its eigenfunctions, yields an expression for $\overline{\overline{\rho}}$ in terms of $\hat{\rho}$:

$$\overline{\overline{\rho}}(y) = \sum_{j=1}^{\infty} \frac{1}{\lambda_j} \langle \hat{\rho}, \psi_j \rangle \mathcal{X}_y(y) \psi_j(y). \qquad (4.3\text{–}8)$$

Equation (4.3–8) shows how it is possible (in principle) to recover $\overline{\overline{\rho}}$ from band-limited data but, because the kernel $\text{sinc}(\Delta k(y - y'))$ is square integrable over $\mathcal{Y} \times \mathcal{Y}$, the solution given by equation (4.3–8) will not be stable. And, while it is possible to regularize this first-kind equation by replacing it with an approximating second-kind equation, it turns out that equation (4.3–1) can be converted into a second-kind equation without the need for any approximation at all.

### 4.3.2  The Papoulis–Gerchberg Algorithm

The band-limited version of equation (4.2–1) is

$$H = F\rho = \mathcal{F}\{\mathcal{X}_y \overline{\overline{\rho}}\} \qquad k_y \in \mathcal{K} \qquad (4.3\text{–}9)$$

where $\mathcal{K} = [2k_1, 2k_2]$.

The adjoint operator $F^\dagger$, defined in appendix definition A.5, is required to obey $\langle F\rho, H \rangle_\mathcal{K} = \langle \rho, F^\dagger H \rangle_\mathcal{Y}$ for all $\rho$ defined on $\mathcal{Y}$ and $H$ defined on $\mathcal{K}$. Substitution of $F\rho$ from equation (4.3–9) yields

$$F^\dagger H = \mathcal{F}^{-1}\{\mathcal{X}_\mathcal{K} H\} \qquad y \in \mathcal{Y}. \qquad (4.3\text{–}10)$$

We can extend equation (4.3–9) to all $k_y \in \mathbb{R}$:

$$F\rho = \mathcal{X}_\mathcal{K} H + (1 - \mathcal{X}_\mathcal{K}) F\rho \qquad k_y \in \mathbb{R}. \qquad (4.3\text{–}11)$$

This is equivalent to

$$\mathcal{F}\{\mathcal{X}_y \overline{\overline{\rho}}\} = \mathcal{X}_\mathcal{K} H + (1 - \mathcal{X}_\mathcal{K}) \mathcal{F}\{\mathcal{X}_y \overline{\overline{\rho}}\} \qquad k_y \in \mathbb{R}. \qquad (4.3\text{–}12)$$

Since equation (4.3–12) is defined for all $k_y \in \mathbb{R}$, it can be inverted using equation (4.3–2) so that

$$\mathcal{X}_y \overline{\overline{\rho}} = \mathcal{F}^{-1}\{\mathcal{X}_\mathcal{K} H\} + \mathcal{F}^{-1}\{(1 - \mathcal{X}_\mathcal{K}) \mathcal{F}\{\mathcal{X}_y \overline{\overline{\rho}}\}\} \qquad y \in \mathbb{R}. \qquad (4.3\text{–}13)$$

Because $\mathcal{X}_y \overline{\overline{\rho}} = \overline{\overline{\rho}}$ when $y \in \mathcal{Y}$, equation (4.3–13) can be written

$$\overline{\overline{\rho}} = \hat{\rho} + K\overline{\overline{\rho}} \qquad y \in \mathcal{Y} \qquad (4.3\text{–}14)$$

where $\hat{\rho} = \mathcal{F}^{-1}\{\mathcal{X}_\mathcal{K} H\}$ and $K\overline{\overline{\rho}} = \mathcal{F}^{-1}\{(1 - \mathcal{X}_\mathcal{K}) \mathcal{F}\{\mathcal{X}_y \overline{\overline{\rho}}\}\}$. This is an integral equation of the second kind and relates the windowed inversion of the data set to the object function $\overline{\overline{\rho}}$. (Equation (4.3–14) should be compared with equation (4.2–9).)

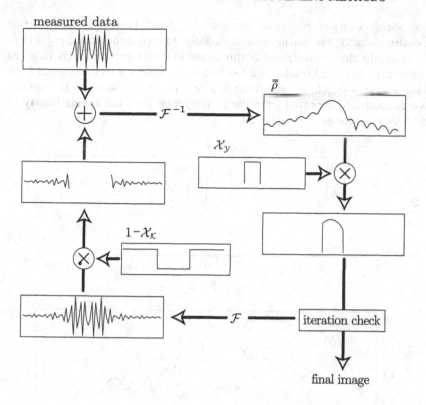

measured data

**Figure 4.2** Flow diagram for the method of alternating orthogonal projections, equation (4.3–15). (Algorithmic flow is clockwise.)

The method of alternating orthogonal projections [17–19] is simply a Neumann iterative solution of this result and can be expressed as

$$\overline{\overline{\rho}}_{est}^{(0)} = 0$$
$$\overline{\overline{\rho}}_{est}^{(j)} = \hat{\rho} + K\overline{\overline{\rho}}_{est}^{(j-1)}, \qquad j = 1, 2, \ldots. \qquad (4.3\text{–}15)$$

(This is illustrated in figure 4.2.)

Figure 4.3 shows how the alternating orthogonal projection algorithm compares with an 'ordinary' estimate $\hat{\rho}$. The actual object is shown in 4.3(a) and was created using $\overline{\overline{\rho}}(y_i) = (y_i - 91)(y_i - 108.5)(y_i - 120)$ on $y_i \in [91, 120]$ and $\overline{\overline{\rho}}(y_i) = 0$ elsewhere (for $y_i = 1, 2, \ldots, 256$). The 256-point DFT was applied to this and the result was truncated so that only the lowest 16 values were retained as 'data' (i.e. support$\{\mathcal{X}_K\} = [1, 2, \ldots, 16]$). Figure 4.3(b) is the simple reconstruction $\hat{\rho}$ determined from these band-limited data. Figure 4.3(c) is the result of 5000 iterations of equation (4.3–15) which were applied using the (known) characteristic function $\mathcal{X}_y$. Note, however, that in actual reconstructions $\mathcal{X}_y$ is usually not accurately known and must be guessed. Moreover, the slow convergence of this example (5000 iterations were required to obtain the displayed results) is a too frequent problem with the original form of the Papoulis–Gerchberg algorithm and recent enhancements to the method have dramatically improved upon the convergence rates for a wide variety of situations (see, for example, [17, 20, 21]).

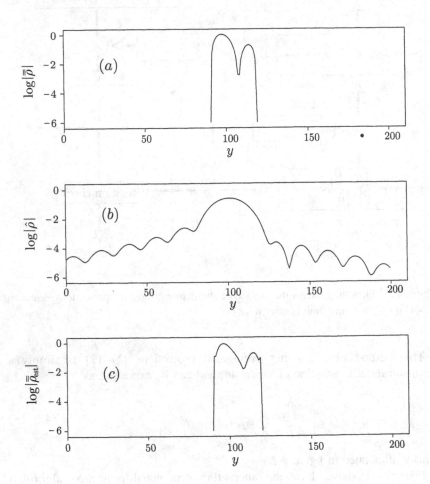

**Figure 4.3**  An example of the alternating orthogonal projections super-resolution method. (*a*) The original object function. (*b*) A simple reconstruction (inverse discrete Fourier transform) based on data consisting of the 256-point discrete Fourier transform of the function with all but the first 16 points discarded. (*c*) The result of 5000 iterations of equation (4.3–15) presuming the actual support of the object function to be known.

The Papoulis–Gerchberg algorithm is recognized to work well provided the support $\mathcal{Y}$ of $\overline{\overline{\rho}}$ is accurately known. Of course, even less prior information was required for the solution of equation (4.3–8) (we assumed only that the target was of finite size so that theorem 4.3.1 applied), but the Papoulis–Gerchberg algorithm is almost always superior. In fact, the addition of some form of *a priori* information turns out to be common to all so-called 'super resolution' methods, and target support is only a (rather crude) version of this.

## 4.4  BAYESIAN METHODS

The two central issues in image reconstruction are limited measurement space and noisy measurements. Up to this point, we have examined methods which attempt to improve the image by including prior null-space information. The stable versions of these imaging methods are 'insensitive' to noise but have made little attempt to account for it explicitly. Bayesian methods can be used to extend the idea of *a priori* knowledge by inserting both null-space and noise information [22, 23].

In the additive noise model of equation (4.2–2), $n$ can be considered to be a random noise vector in the sense that each measurement will generally contain a different realization of $n$ selected from an ensemble of all possible noise vectors. This noise ensemble can be described by a probability density function $p_N$ which we shall take to be a multivariate Gaussian with zero mean:

$$p_N(n) = \frac{1}{C(\det R_n)^{1/2}} \exp\left(-\tfrac{1}{2}n^\dagger R_n^{-1} n\right) \qquad (4.4\text{--}1)$$

where $R_n$ is the noise covariance matrix and $C$ is a normalizing constant.

Because of equation (4.2–2), these data will also be a realization of an ensemble and, consequently, so will any estimated object function based on this data set. These ensembles are induced by the noise ensemble *and any additional* uncertainties that we may harbor about the data set, or $\rho$, or both, and are described by probability density functions of their own. *Bayes formula* expresses the *a posteriori* conditional probability of $\rho$ given $H$ as

$$p_{P|H}(\rho) = \frac{p_{H|P}(H)p_P(\rho)}{p_H(H)}. \qquad (4.4\text{--}2)$$

In this equation, $p_{H|P}$ is the probability density of $H$ given $\rho$, and $p_P$ and $p_H$ are the *a priori* densities of $\rho$ and $H$, respectively.

From equation (4.4–1) we can conclude that the conditional density of these data $H$ (given the object function $\rho$) is determined by the noise statistics so that

$$p_{H|P}(H) = p_N(H - F\rho). \qquad (4.4\text{--}3)$$

The *a priori* probability density $p_P$ permits the inclusion of null-space information in the form of 'possible attributes' of the target along with their probabilities. For reasons of mathematical tractability, this function is often modeled as

$$p_P(\rho) = \frac{1}{C(\det R_\rho)^{1/2}} \exp\left[-\tfrac{1}{2}(\rho - \rho_0)^\dagger R_\rho^{-1}(\rho - \rho_0)\right] \tag{4.4-4}$$

where $\rho_0$ is the mean value of $\rho$ in this Gaussian distribution and $R_\rho$ is the covariance matrix. Both $\rho_0$ and $R_\rho$ are considered to be model parameters. (The *a priori* density $p_H$ is, for our concerns, effectively just a normalization coefficient and will not be discussed further.)

When the model parameters are specified, equation (4.4–2) yields the probability density $p_{P|H}(\rho)$ which describes the ensemble of all solutions consistent with the noise and the prior image model (possibly containing null-space component assumptions). Substituting equations (4.4–3) and (4.4–4) into equation (4.4–2) yields

$$p_{P|H}(\rho) \sim \exp\left\{-\tfrac{1}{2}\left[(H - F\rho)^\dagger R_n^{-1}(H - F\rho) + (\rho - \rho_0)^\dagger R_\rho^{-1}(\rho - \rho_0)\right]\right\}. \tag{4.4-5}$$

After some algebraic maneuvering, this can be written in the form

$$p_{P|H}(\rho) \sim \exp\left[-\tfrac{1}{2}(\rho - \langle\rho\rangle)^\dagger (R_\rho^{-1} + F^\dagger R_n^{-1}F)(\rho - \langle\rho\rangle)\right] \tag{4.4-6}$$

where $\langle\rho\rangle$ obeys

$$(R_\rho^{-1} + F^\dagger R_n^{-1}F)(\langle\rho\rangle - \rho_0) = F^\dagger R_n^{-1}(H - F\rho_0). \tag{4.4-7}$$

The image

$$\rho_{\text{MAP}} = \arg\max_{\rho \in L^2(\mathbb{R})} \{p_{P|H}(\rho)\} \tag{4.4-8}$$

is known as the *maximum a posteriori* estimate. This estimate will be the one for which the argument of the exponential in equation (4.4–6) vanishes and, consequently, equation (4.4–7) yields the MAP solution as $\rho_{\text{MAP}} = \langle\rho\rangle$. Multiplying equation (4.4–7) by $R_\rho$ and simplifying yields

$$\rho_{\text{MAP}} = \rho_0 + R_\rho F^\dagger R_n^{-1}(H - F\rho_{\text{MAP}}). \tag{4.4-9}$$

Equation (4.4–9) is a second-kind equation and can be solved iteratively. Note, however, that it is essential to specify $\rho_0$ and $R_\rho$ in a non-trivial way: simply setting $\rho_0 = $ constant and $R_\rho = \sigma^{-2}I$, for example, does nothing to quantify the null-space components of $H = F\rho$ and, consequently, cannot mitigate the effects of limited data. (Note that when $F$, $R_\rho$ and $R_n$ are stationary, then the solution to equation (4.4–9) is essentially the same as that given by the Wiener filter [12].)

When no prior information is known about $\rho$ then we can represent this lack of an *a priori* image model by setting $p_P$ to a uniform density function. In this case, the MAP solution will be the same as the one determined by the *minimum chi-squared error method*: $\rho_{\chi^2} = \arg\min\{\chi^2\}$ where

$$\chi^2(\rho) = (H - F\rho)^\dagger R_n^{-1}(H - F\rho). \tag{4.4-10}$$

($\rho_{\chi^2}$ is also known as the *maximum likelihood* solution.)

If we compare the argument of the exponential in equation (4.4–5) with equation (4.2–12), we can see that the methods are formally the same when $\alpha = 1$, $\mathcal{J}_1(\rho, h) = \chi^2(\rho)$, and $\mathcal{J}_2(\rho, \rho_0) = (\rho - \rho_0)^\dagger R_\rho^{-1}(\rho - \rho_0)$. Under this interpretation, the functional $\mathcal{J}_1$ will be determined by the noise statistics (which, of course, need not be Gaussian). The choice for $\mathcal{J}_2$, which measures some kind of 'distance' between our estimated solution and our *a priori* information, will depend on just what it is that we know about the target and how we want to model it.

Equation (4.4–4) is a convenient choice because it leads to the linear equation of (4.4–9), but this choice is not always a good representation for a target model and others are often made. Maximum entropy reconstruction, for example, sets [24, 25]

$$p_P(\rho) \sim \exp\left[ -\frac{\alpha}{2} \int_y p_m(\overline{\overline{\rho}}(y')) \ln\left( \frac{p_m(\overline{\overline{\rho}}(y'))}{p_0(y')} \right) dy' \right]. \tag{4.4-11}$$

The function $p_m(\overline{\overline{\rho}}(y))$ is an additional density function representing our model. The argument of the exponential in equation (4.4–11) is proportional to the 'cross-entropy' and measures the *information difference* between the model density $p_m(\overline{\overline{\rho}}(y))$ and the prior image density $p_0(y)$—it is the expectation of the difference between $\ln p_m(\overline{\overline{\rho}})$ and $\ln p_0$. The functional form of $p_m(\cdot)$, of course, is pre-specified as part of the model and so the density $p_P(\rho)$ will measure the probability of the information difference between an image realization $\overline{\overline{\rho}}(y)$ and the prior image (embodied in $p_0(y)$). In practice, $p_m$ is difficult to estimate and the most common image model sets $p_0(y)$ to a uniform density and $p_m(\overline{\overline{\rho}}(y)) \propto \overline{\overline{\rho}}(y)$—the so-called *configuration density*.

The configuration density model assumes $\overline{\overline{\rho}}(y) > 0$ for all $y$ and is inappropriate for the complex-valued images often used in radar. Various extensions to $p_m$ have been proposed, such as $p_m(\overline{\overline{\rho}}(y)) \propto |\overline{\rho}(y)|$, but these have not been universally accepted. Determination of the 'correct' form for $p_m(\overline{\overline{\rho}}(y))$ is still a problem and has caused some investigators to question the applicability of entropy-based methods to radar imaging problems.

## 4.5   MODEL-BASED RESOLUTION IMPROVEMENT

The three preceding resolution enhancement techniques are not extensively used because they are often more trouble than they are worth: the resulting image improvement is often inconsequential while the computational cost may be quite significant. There is, however, a commonly used method that, when appropriate, can dramatically improve the image quality and requires (relatively) small computational expense. As with the other methods examined, this approach relies on the properties of an image model but gains its usefulness over the others by employing an especially simple representation that is particularly appropriate to radar imaging.

In the introduction to chapter 3 we noted that the inverse problem could be attacked by developing a parametric image model and using the direct scattering problem to predict the associated (model) scattered field. In this approach, the estimated image will be determined by the set of parameters that minimize the difference between the measured data and the modeled field (for example by least-squares).

One of the main problems with this kind of solution comes out in its implementation—it is one thing to talk about minimizing a difference functional, but developing an efficient algorithm that puts it into effect is often another thing altogether. (Problems in developing effective search strategies which must account for local minima can be quite formidable.) One common and simple search scheme that can be applied to the radar imaging problem in an efficient manner is known as the CLEAN algorithm [26]. Originally devised by radio astronomers in the mid-1970s to mitigate sidelobes caused by limited data, the CLEAN approach uses a point scatterer model (and the consequently simple image model) to significantly improve image quality. This image model is of the form

$$\overline{\overline{\rho}}_{\text{model}}(y) = \sum_{n=1}^{N} A_n \delta(y - y_n), \qquad A_n \in \mathbb{C} \qquad (4.5\text{--}1)$$

so that

$$\mathcal{F}\{\overline{\overline{\rho}}_{\text{model}}\}(k) = \frac{1}{2\pi} \sum_{n=1}^{N} A_n e^{i2ky_n} \qquad (4.5\text{--}2)$$

and the basic idea is to compare a (low-resolution) reconstructed image element obtained from the measured data with its idealized image representation and the associated idealized data.

The low-resolution image used by the CLEAN algorithm is just the windowed Fourier transform: $\hat{\rho} = \mathcal{F}^{-1}\{\mathcal{X}_K H_K\}$. This $\hat{\rho}$ will be contaminated by sidelobes of the sinc$(y)$ function (see equation (4.3–3)), but when the noise level is small, the scatterer model of equation (4.5–1) ensures that the location and strength of the largest peak will (usually) coincide with the location and strength of its associated point scatterer. This property forms the basis of the CLEAN method and the idealized image is (iteratively) modified to include a point scatterer (of appropriate strength) at the location indicated by a peak in the low-resolution image. The effects of this idealized image element are then removed from the data by *subtracting* the idealized band-limited components of field calculated in

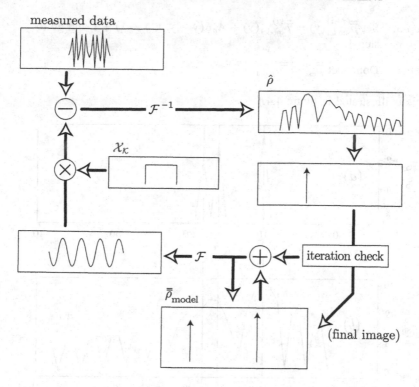

**Figure 4.4**  Flow diagram for the CLEAN algorithm. (Algorithmic flow is clockwise.)

the direct problem (using this single point scatterer). This subtraction is 'allowed' because the target model consists of non-interacting elements and it is simple and quick because the elements are points.

The data subtraction step essentially 'cancels' the effects that the idealized target element has on the measured field while retaining the position and strength information in the ideal image. (Note that since the data to be subtracted were calculated as a direct scattering problem, this cancelation includes the associated sidelobes as well.) Since the central peak of the $\mathrm{sinc}(y)$ function is larger than its sidelobes, the algorithm is performed successively (in order of lessening scatterer strength) for all such peaks and can be detailed as follows:

step 0:  Set the iteration index $j = 0$ and $\overline{\overline{\rho}}_{\mathrm{model}}^{(j)}(y) = 0$ for all $y$

step 1:  Calculate $\hat{\rho} = \mathcal{F}^{-1}\{\mathcal{X}_K(H_K - \mathcal{F}\{\overline{\overline{\rho}}_{\mathrm{model}}^{(j)}\})\}$

step 2:  Find $y_n = \arg\max\{|\hat{\rho}(y)|\}_y$ and set $A_n = \hat{\rho}(y_n)$. If $|A_n| <$ threshold, go to step 4

step 3:   Set $\overline{\overline{\rho}}_{\text{model}}^{(j+1)}(y) = \overline{\overline{\rho}}_{\text{model}}^{(j)}(y) + A_n \delta(y - y_n)$, $j \to j + 1$ and go to
         step 1

step 4:   Done; set $\overline{\overline{\rho}}_{\text{est}} = \overline{\overline{\rho}}_{\text{model}}$.

(This is illustrated in figure 4.4.)

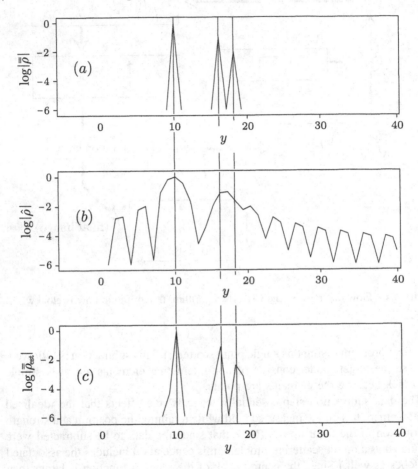

**Figure 4.5** Example CLEAN-based image. (a) Shows the original (synthetic) target. (b) A simple reconstruction (inverse discrete Fourier transform) based on data consisting of the 96-point discrete Fourier transform of the function with all but the first 32 points discarded. (c) The result of five iterations of the CLEAN algorithm.

Figure 4.5 displays a CLEAN reconstruction for an ideal test target consisting of three point scatterers. This synthetic target is shown in 4.5(a) and was created using $\overline{\overline{\rho}}(y_{10}) = 1$, $\overline{\overline{\rho}}(y_{16}) = 0.5$, and $\overline{\overline{\rho}}(y_{18}) = 0.25$ (and zero elsewhere). The 96-point DFT was applied to this target and truncated so that only the low 32 values were retained as 'data'. Figure 4.5($b$) is the simple reconstruction $\hat{\rho}$ determined from these band-limited data. Figure 4.5($c$) is the result of five iterations of the CLEAN algorithm.

Because the CLEAN technique relies on the low-resolution (windowed Fourier transform) image to estimate the location of the idealized image elements, it is often not considered to be a 'super-resolution' method. Two scattering centers which lie within the central lobe of the sinc($y$) function will yield a low-resolution image that appears as a single peak and the CLEAN method may treat them as a single idealized scatterer. The resulting calculated scattered field will then differ from the measured data and, consequently, the cancelation step (step (1)) will not always be complete—possibly resulting in image artifacts. For this reason, the CLEAN approach is sometimes combined with other methods (for example Papoulis–Gerchberg) in an effort to both 'clean' and super-resolve the image estimate.

# REFERENCES

[1]   Das Y and Boerner W-M 1978 On radar target shape estimation using algorithms for reconstruction from projections *IEEE Trans. Antennas Propag.* **26** 274

[2]   Mager R D and Bleistein N 1978 An examination of the limited aperture problem of physical optics inverse scattering *IEEE Trans. Antennas Propag.* **26** 695

[3]   Gross F B and Young J D 1981 Physical optics imaging with limited aperture data *IEEE Trans. Antennas Propag.* **29** 332

[4]   Bennett C L 1981 Time domain inverse scattering *IEEE Trans. Antennas Propag.* **29** 213

[5]   Noel B 1991 *Ultra-Wideband Radar, Proc. First Los Alamos Symp.* ed B Noel (New York: CRC Press)

[6]   Hudson S and Psaltis D 1993 Correlation filters for aircraft identification from radar range profiles *IEEE Trans. Aerospace Electr. Syst.* **29** 741

[7]   Adachi S and Uno T 1993 One-dimensional target profiling by electromagnetic backscattering *J. Electromag. Waves Appl.* **7** 403

[8]   Engl H W, Hanke M and Neubauer A 1996 *Regularization of Inverse Problems (Mathematics and Its Applications Series)* vol 375 (New York: Kluwer Academic)

[9]   Rushforth C K 1987 Signal restoration, functional analysis, and Fredholm integral equations of the first kind *Image Recovery: Theory and Application* ed H Stark (Orlando, FL: Academic)

[10]  Hutson V and Pym J S 1980 *Applications of Functional Analysis and Operator Theory* (New York: Academic)

[11]  Nashed M Z 1981 Operator-theoretic and computational approaches to ill-posed problems with applications to antenna theory *IEEE Trans. Antennas Propag.* **29** 220

[12]  Andrews H C and Hunt B R 1977 *Digital Image Restoration* (Englewood Cliffs, NJ: Prentice-Hall)

[13] Sage A P and Melsa J L 1979 *Estimation Theory with Applications to Communications and Control* (New York: Krieger)

[14] Harris F J 1978 On the use of windows for harmonic analysis with the discrete Fourier transform *Proc. IEEE* **66** 1

[15] Slepian D and Pollak H O 1961 Prolate spheroidal wave functions, Fourier analysis and uncertainty – I *Bell Syst. Tech. J.* **40** 43

[16] Flammer C 1957 *Spheroidal Wave Functions* (Stanford, CA: Stanford University Press)

[17] Rhebergen J B, van den Berg P M and Habashy T M 1997 Iterative reconstruction of images from incomplete spectral data *Inverse Problems* **13** 829

[18] Gerchberg R W 1979 Super-resolution through error energy reduction *Opt. Acta* **14** 709

[19] Papoulis A 1975 A new algorithm in spectral analysis and band-limited extrapolation *IEEE Trans. Circuits Syst.* **22** 735

[20] Sanz J L C and Huang T S 1983 Unified Hilbert space approach to iterative least-squares linear signal restoration *J. Opt. Soc. Am.* **73** 1455

[21] Maitre H 1981 Iterative superresolution: some new fast methods *Optica Acta* **28** 973

[22] Hunt B R 1977 Bayesian methods in nonlinear digital image restoration *IEEE Trans. Comput.* **26** 219

[23] Herman G T and Lent A 1976 A computer implementation of a Bayesian analysis of image reconstruction *Inform. Contr.* **31** 364

[24] Mohammad-Djafari A and Demoment G 1989 Maximum entropy Fourier synthesis with application to diffraction tomography *Appl. Opt.* **26** 1745

[25] Frieden B R and Bajkova A T 1994 Bayesian cross-entropy of complex images *Appl. Opt.* **33** 219

[26] Tsao J and Steinberg B D 1988 Reduction of sidelobe and speckle artifacts in microwave imaging: the CLEAN technique *IEEE Trans. Antennas Propag.* **36** 543

# 5

---

# Two-Dimensional Imaging

It may be possible to use the one-dimensional images determined by equation (4.1–2) as the basis for a target recognition scheme, and some of these efforts are referenced at the end of chapter 4. Be that as it may, HRR profiles are fundamentally limited because they do not reveal any information about the location of target components in directions orthogonal to $\hat{R}$ (this is obvious from equations (4.1–1) and (4.1–2)). As in optical imaging systems, such cross-range information is available only when radar data are collected over a non-zero aperture $A$. Unlike optical systems, radar waveforms are relatively insensitive to atmospheric attenuation because they employ significantly longer wavelengths $\lambda = 2\pi/k$. Unfortunately, these longer-wavelength radar systems are constrained by a resolution limit ratio $\lambda/A$ that is correspondingly large. Consequently, deployable fixed-aperture systems using these longer wavelengths will suffer from poor image cross-range resolution and, as a result, 'ordinary' radar methods can never be expected to achieve the same level of resolution as is typically seen in even 'small' optical systems. If, however, these data are collected from an *extended* aperture which is not fixed to the physical dimensions of the radar antenna, then the ratio $\lambda/A$ can (in principle) be made small enough to yield optical quality images.

This is the basic idea behind synthetic aperture radar: data are collected from many locations $\{x_i\}$ (distributed over a region much larger than the radar antenna) by physically shifting the position of the radar relative to that of the target. The terminology is somewhat ambiguous in describing these methods, but usually SAR (synthetic aperture radar) refers to a moving system which collects data from a stationary target, while ISAR (inverse SAR) implies a fixed system collecting data from a rotating target. The basic data processing issues are essentially the same for both situations, although there are some obvious practical differences. In problems of aircraft target identification we are almost always concerned with ISAR data and their interpretation.

The problems associated with image estimation in one dimension are fundamentally unaltered in two dimensions, and most of the results established in sections 4.2–4.6 carry over with only minor alteration. In addition to ill-posedness and resolution enhancement, however, there are three important issues *unique* to ISAR-based two-dimensional imaging, and they will be examined in this chapter. The first is a result of the way in which these data are collected and the need to accurately map these data to the positions $\{x_i\}$ at which they are measured. The second problem is a consequence of the 'phase-coherent' nature of the complex-valued scattered field measurements. Third, and perhaps most difficult, is the fact that the simple weak scatterer model implied by equation (3.4–3) is not completely accurate and its straightforward application results in image artifacts that can greatly complicate the interpretation of an ISAR image. The first of these issues will be discussed here, while the last two problems (scintillation and scattering model error) will be deferred to chapter 6.

## 5.1  THE BASIC IMAGING EQUATION

When $\theta$ is allowed to vary during the data collection process, equation (4.1–1) must be generalized. If we define a new set of coordinates $(u, v)$ by

$$u \equiv x \cos\theta + y \sin\theta$$
$$v \equiv y \cos\theta - x \sin\theta \qquad (5.1\text{–}1)$$

(see figure 5.1), then we can write

$$\overline{\overline{\rho}}_\theta(v) = S\{\overline{\rho}\}(v, \theta) \equiv \frac{1}{2\pi} \int_{L(v;\theta)} \overline{\rho}(x', y') \, du'$$

$$= \frac{1}{(2\pi)^2} \int_{L(v;\theta)} \int_{\mathbb{R}} \rho(x', y', z') \, dz' \, du' \qquad (5.1\text{–}2)$$

where $L(v; \theta) = \{(x, y)|y \cos\theta - x \sin\theta = v\}$.

The range profile $\overline{\overline{\rho}}_\theta$ is similar to the 'shadow transform' of $\overline{\rho}$ used in medical tomographic imaging, except that here it is projected onto the range axis. Using this notation, the so-called 'projection slice' theorem is just a generalization of equation (4.1–2):

$$H(k, \theta) = \frac{1}{2\pi} \int_{\mathbb{R}} \overline{\overline{\rho}}_\theta(v') e^{i2kv'} \, dv', \quad k \in [k_1, k_2]. \qquad (5.1\text{–}3)$$

If we make the change of variables $\xi = -2k \sin\theta$, $\zeta = 2k \cos\theta$, so that $\partial(\xi, \zeta)/\partial(k, \theta) = 4|k|$, then equation (3.4–3) can be used to obtain

$$\hat{\rho}(x, y) = \int_{\mathbb{R}^2} H(k', \theta') e^{-i(\xi'x + \zeta'y)} \, d\xi' \, d\zeta'$$

$$= \int_0^\pi d\theta' \int_{-\Delta k/2}^{\Delta k/2} H(k', \theta') e^{-i2k'(y \cos\theta' - x \sin\theta')} 4|k'| \, dk' \qquad (5.1\text{–}4)$$

where $\Delta k = k_2 - k_1$ denotes the bandwidth (the range of $k$ values).

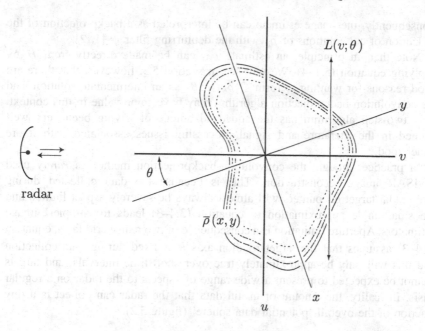

**Figure 5.1** Geometry for calculating the range profile $\overline{\overline{\rho}}_\theta$ appropriate to ISAR data.

The *convolution-backprojection* algorithm is a two-step method for obtaining an estimate $\rho_{est}$ of $\overline{\rho}(x, y)$ from the shadow function $\overline{\overline{\rho}}_\theta(k)$. The inner integral in equation (5.1–4) can be written

$$\mathcal{Q}\{\overline{\overline{\rho}}_\theta\}(v; \theta) = \int_{-\Delta k/2}^{\Delta k/2} H(k', \theta) e^{-i2k'(y\cos\theta - x\sin\theta)} 4|k'|\, dk'$$

$$= \frac{1}{2\pi} \int_{-\Delta k/2}^{\Delta k/2} \left( \int_{\mathbb{R}} \overline{\overline{\rho}}_\theta(v') e^{i2k'v'}\, dv' \right) e^{-i2k'v} 4|k'|\, dk' \tag{5.1–5}$$

and can be seen to be a convolution of $\overline{\overline{\rho}}_\theta$ with the 'deblurring filter'

$$\psi(v) = \int_{-\Delta k/2}^{\Delta k/2} 4|k'| e^{i2k'v}\, dk' \tag{5.1–6}$$

$$= 2\Delta k^2 \operatorname{sinc}(\Delta k v) - \Delta k^3 \operatorname{sinc}^2(\tfrac{1}{2}\Delta k v).$$

Let $q(v; \theta) = \mathcal{Q}\{\overline{\overline{\rho}}_\theta\}(v; \theta)$ denote the *filtered projection at angle* $\theta$. The outer integral in equation (5.1–4) defines the so-called backprojection operator as

$$\mathcal{B}\{q\}(x, y) = \int_0^\pi q(y\cos\theta' - x\sin\theta'; \theta')\, d\theta'. \tag{5.1–7}$$

Consequently, the image estimate can be interpreted as a backprojection of the collection of convolutions of $\overline{\overline{\rho}}_\theta$ with the deblurring filter $\psi$ [1, 2].

Note that, in principle, an estimate $\rho_{est}$ can be made directly from $H$ by applying equation (5.1–4). We will see in section 5.3, however, that there are good reasons for wanting to form $\mathcal{Q}\{\overline{\overline{\rho}}_\theta\}(v; \theta)$ as an intermediate solution, and the convolution-backprojection algorithm may have some value in this context. This two-step algorithm has the added advantage of having been very well studied in the literature and signal processing issues associated with it are understood.

In practice, though, the convolution-backprojection method is *rarely* used in ISAR image reconstruction. This is because the data collected during most radar/target encounters will almost always be severely aspect-limited and the small-angle approximation to equation (3.4–3) leads to 'simpler' image estimators. Aperture limitation is a consequence of two real-world facts: equation (3.4–3) assumes that the target rotation axis $\hat{k}$ is fixed during data collection and this will only be approximately true over short time intervals; and targets cannot be expected to present a wide range of aspects to the radar on a regular basis. In reality, the amount of useful data that the radar can collect is a tiny fraction of the overall 'potential data sphere' (figure 5.2).

## 5.2  DATA ERRORS

Usually, most ISAR image estimates of airborne targets rely on the small-aperture approximation [3–22]. If we let $\sin\theta \approx \theta$ and $\cos\theta \approx 1$, then equation (3.4–3) becomes

$$H(k, \theta) \approx \frac{1}{(2\pi)^2} \int_{\mathbb{R}^2} \overline{\rho}(x', y') e^{-i(k_x x' + k_y y')} \, dx' \, dy' \qquad (5.2–1)$$

where now, $k_x = 2k\theta$ and $k_y = -2k$. Equation (5.2–1) is sometimes called the standard ISAR image equation and the small-aperture approximation allows us to identify (up to a translation) the target-centric coordinates with the radar-centric coordinates. The integrand in equation (5.2–1) divides the target into down-range and cross-range factors and it is easy to visualize ISAR imaging in these terms.

If we multiply equation (5.2–1) by $\exp[i(k_x x + k_y y)]$ and integrate over all $k_x$ and $k_y$ appropriate to these data, then we can form an estimate $\hat{\rho}(x, y)$ (in terms of $\overline{\rho}(x, y)$) as

$$\hat{\rho}(x, y) = \overline{k}\Delta\theta \Delta k \int_{\mathbb{R}^2} \overline{\rho}(x', y') \\ \times \text{sinc}[\overline{k}\Delta\theta(x - x')] \, \text{sinc}[\Delta k(y - y')] \, dx' \, dy' \qquad (5.2–2)$$

where $\Delta\theta = \theta_2 - \theta_1$ denotes the aperture (range of $\theta$ values), and $\overline{k} = (k_1 + k_2)/2$.

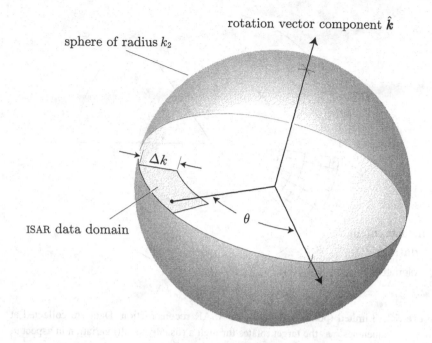

rotation vector component $\hat{k}$

sphere of radius $k_2$

$\Delta k$

$\theta$

ISAR data domain

**Figure 5.2** ISAR data domain. $\Delta k$ is the bandwidth, $k_2$ is the largest (spatial) frequency measured, and $\theta$ is the rotation coordinate representing target aspect.

The quality of the target reconstruction will be determined by the size of the target resolution cell whose down-range and cross-range dimensions are inversely proportional to the bandwidth and aperture, respectively (specifically, $\Delta x = 2\pi(\bar{k}\Delta\theta)^{-1}$ and $\Delta y = 2\pi\Delta k^{-1}$). Note, however, that in practice $\theta$ is determined from an estimate of the target's rotation rate $\Omega$. When this rate is unknown, the cross-range dimension is usually expressed in terms of Doppler frequency shift which, for a rotating rigid target, will be proportional to the cross-range position of a target's scattering sub-element. If $\Delta f_D$ denotes the Doppler frequency resolution, then the cross-range resolution can be expressed as $\Delta x = 2\pi\Delta f_D(\bar{k}\Omega)^{-1}$. There is, of course, a trade-off between cross-range and down-range resolution which is determined by the ambiguity function of equation (2.6–1).

### 5.2.1   Polar Reformatting

Denote by **H** (with $\mathsf{H}_{ij} \equiv H(k_i, \theta_j)$) the measured complex data array, indexed by $k$ and $\theta$. The rows of **H** correspond to different frequencies of the scattered field while the columns correspond to different aspects. The small-angle approximation of equation (5.2–1) introduces a source of phase errors in

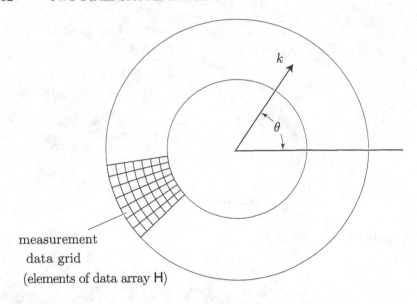

measurement
data grid
(elements of data array H)

**Figure 5.3** Limited (polar) data space for ISAR reconstruction. Data are collected at multiple frequencies $k$ as the target rotates through a (usually small) variation in aspect $\theta$. (See figure 5.2.)

H since, while these data are collected on a polar grid (they are said to be in *polar format*), the estimate obtained from equation (5.2–1) assumes this grid to be *rectangular* (they are assumed to be in *frequency space*) [23, 24]. If the bandwidth is small and the aperture is sufficiently narrow, then the polar grid is approximately square and the reconstruction based on equation (5.2–1) will be sufficient for many applications (this can be seen by inspecting figure 5.3). In high-resolution imaging situations, however, estimates based on equation (5.2–1) can result in blurring artifacts—especially when the support of $\overline{\rho}(x, y)$ contains large values of $x$ or $y$. If we refine the small-angle approximation so that $\sin \theta \approx \theta - \frac{1}{6}\theta^3$ and $\cos \theta \approx 1 - \frac{1}{2}\theta^2$ then, instead of equation (5.2–1), equation (3.4–3) can be approximated as

$$
H(k, \theta) \approx \frac{1}{(2\pi)^2} \int_{\mathbb{R}^2} \int_{\mathbb{R}^2} \overline{\rho}(x', y') B(x', y', x'', y'')
$$
$$
\times e^{-i(k_x x'' + k_y y'')} \, dx' \, dy' \, dx'' \, dy'' \tag{5.2–3}
$$

where

$$
B(x', y', x'', y'') = \int_{\mathbb{K}} \exp\left(i\frac{k_x'^3}{6k_y'^2}x' + i\frac{k_x'^2}{2k_y'}y'\right)
$$
$$
\times e^{i(k_x'(x''-x') + k_y'(y''-y'))} \, dk_x' \, dk_y' \tag{5.2–4}
$$

and the region $\mathbb{K}$ is defined by the data support of figure 5.3. Equation (5.2–3) shows that these blurring artifacts can be understood as a consequence of the spatially varying point-spread function of equation (5.2–4) [32].

Deconvolution of equation (5.2–3) will correct for these errors, but the standard method is to interpolate the measured data and resample on the uniform square grid implied in equation (5.2–1). This process, which is also sometimes called 'focusing', is illustrated by figure 5.4.

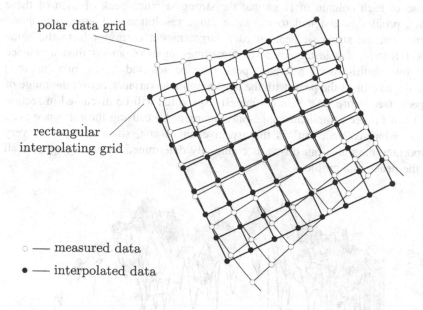

**Figure 5.4**  Polar reformatting of these data is accomplished by interpolation of the original (polar) data and resampling it on a rectangular grid. When $\Delta\theta$ and $\Delta k$ are sufficiently small, this step is often omitted.

### 5.2.2  Motion Compensation

Another source of phase errors across **H** can be interpreted as a result of equation (3.4–2) which was formally designed to 'absorb' the complex range-dependent prefactor $ikH_0R^{-1}\exp\left[i\left(2kR-\omega t\right)\right]$ of equation (3.4–1) into these data. Up to this point of the discussion we have tacitly assumed that this complex prefactor can be divided through before the integral transform is 'inverted' (equation (5.2–2)). But since these data are collected from a series of radar pulses scattered from a target as it moves in space, there will usually be a variation in the range $R$ over the data set and, effectively, $R = R(\theta)$. This variation is typically too small to be of consequence in assigning the magnitude of the prefactor but, because of the large values of $k$ we are considering, the range phaseshift can be quite significant [25–31].

This problem of 'range walking' results in the introduction of phase errors across the data set and $R(\theta)$ needs to be estimated so that these data can be carefully 'range-bin aligned' before the inversion is performed. The Fourier transform of each column of $\mathbf{H}$ will be a range profile of the form of equation (5.1–2). Range-bin alignment can effectively be accomplished by adjusting the phase of each column of $\mathbf{H}$ so that the strongest range-peak of each of these range profiles is assigned to the same range resolution cell—the assumption being that the strongest peak at each target aspect corresponds to the same target feature. Range alignment is sometimes more of an art than a science and this (multi-aspect) reference peak may be selected as the first (nearest) strong peak, or as the peak with the least amplitude variance across the range of aspects (see figure 5.5). Scintillation effects (which will be discussed in section 6.2) can further complicate range-bin alignment by causing the reference peak to occasionally 'fade-out' as the aspect changes and, since accuracy is very important, the range-bin off-set is commonly determined by averaging over all of the peaks in a profile.

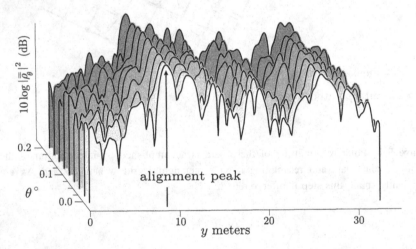

**Figure 5.5** Range profile alignment. A large peak is selected and assumed to represent the same scattering center in each profile $\overline{\overline{\rho}}_\theta$. Range offset is calculated and each profile is phase compensated by aligning the peaks.

Define a correlation function between the two range profiles $\overline{\overline{\rho}}_{\theta_i}(v)$ and $\overline{\overline{\rho}}_{\theta_j}(v)$, corresponding to two aspects $\theta_i$ and $\theta_j$, by

$$A_{ij}(u) = \int_{\mathbb{R}} |\overline{\overline{\rho}}_{\theta_i}(v')||\overline{\overline{\rho}}_{\theta_j}(v'+u)| \, \mathrm{d}v' . \qquad (5.2\text{--}5)$$

Denote by $\delta R(i, j)$ the range correction required to 'align' $\overline{\overline{\rho}}_{\theta_i}$ and $\overline{\overline{\rho}}_{\theta_j}$. If we assume $\overline{\overline{\rho}}_{\theta_i}(v + \delta R(i, j)) \approx \overline{\overline{\rho}}_{\theta_j}(v)$ then

$$\delta R(i, j) = \arg \max_{u \in \mathbb{R}} A_{ij}(u). \tag{5.2-6}$$

Let $\Delta R_j = \delta R(j - 1, j)$ denote the range shift required to correct the $j$th range profile. Then equation (5.2–6) provides a prescription for adjusting the range error: starting with $j = 2$, sequentially step through the data array adjusting each element by $H_{ij} \rightarrow H_{ij} \exp(-i2k_i \Delta R_j)$.

### 5.2.3  Example ISAR Image

In practice, these methods for correcting polar-format-induced defocusing and target range-walk are often *not* performed. Polar reformatting will only be required when the aperture is larger than the small-angle approximation and/or the target is very large. Motion compensation, on the other hand, will usually be required—but when the target moves uniformly, the motion-induced phaseshift can often be corrected by fitting an estimated target motion curve to the observed target path. This involves accurate tracking of the target's range by (typically) timing the radar pulse return. For example, current radar systems offer sufficient accuracy to estimate target position, velocity, acceleration and jerk, and it is possible to fit a third-order polynomial to the target's radial path:

$$R(t) \approx R(0) + \dot{R}(0)t + \tfrac{1}{2}\ddot{R}(0)t^2 + \tfrac{1}{6}\dddot{R}(0)t^3 \tag{5.2-7}$$

where (`) denotes differentiation with respect to time. (Note, for accuracy, that in the target-centric coordinate system we are using, the motion is technically that of the radar. Since it is actually relative motion between the radar and the target that is in play here, equation (5.2–7) should cause no confusion.)

Figure 5.6 is an example ISAR image created with data collected from a B-727 jetliner as it was taking off [32]. The target was located approximately three miles from the radar and was undergoing a steady turn while the radar measurements were made. Motion compensation was accomplished by fitting the third-order polynomial of equation (5.2–7) to the target's trajectory and using this polynomial to determine the pulse-to-pulse phase correction $\Delta R_j$. (Figure 5.5 uses this same data and illustrates the accuracy of the motion compensation.) The radar's center frequency was $\overline{\omega} = 2\pi \times 9.25$ GHz and its bandwidth was $\Delta \omega = 2\pi \times 500$ MHz. These data were collected over a short time (several seconds) and resulted in an overall aperture of $\Delta \theta \approx 2°$.

## 5.3  RESOLUTION IMPROVEMENT

Almost all of the results from sections 4.4 and 4.5 can be applied to two-dimensional radar imaging. Efficient implementation of these algorithms,

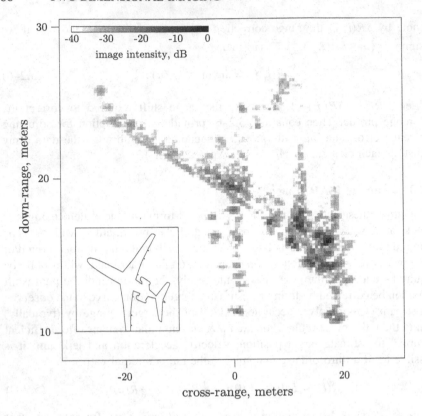

**Figure 5.6** Example ISAR image. The target is a B-727 jetliner with orientation as in inset. The radar center frequency and bandwidth were 9.25 GHz and 500 MHz, respectively. Compare this figure with the range profile of figure 4.1.

however, typically requires more sophisticated methods than Neumann iteration (the CLEAN algorithm is an obvious exception to this, of course, and the algorithm of section 4.5 has an easy extension to two dimensions [33]). The Papoulis–Gerchberg algorithm can also be extended as a Neumann iterative solution to the two-dimensional ISAR imaging equation, but this is often not a wise choice because of the computational requirements associated with each iteration.

Because equation (5.2–1) can be written as an iterated integral over the cross-range and down-range variables, the simplest approach is to tackle each coordinate direction separately so that, for example, the $\overline{\rho}_{x\ \text{est}}(x, y_i)$ are determined for each $y_i$ by treating $x$ as a variable and $y_i$ as a fixed parameter, and then reversing directions to determine $\overline{\rho}_{yx\ \text{est}}(x_i, y)$ by treating $y$ as the variable and $x_i$ as the parameter. This approach is similar to the non-uniform windowing methods often used in two-dimensional DFT calculations which are modeled as the product of two one-dimensional windows. It is easy to see, however, that this 'simple' approach can become a computational nightmare since, if there are $M$ cross-range bins, $N$ down-range bins and $L$ (Neumann) iterations per subestimate, then there will be $M \times N \times L$ iterations required for the final estimate of $\overline{\rho}(x, y)$.

A more practical approach attacks the integral equation (4.4–14) directly, and there are a variety of iterative methods that converge faster than the simple Neumann scheme (cf [34] and references cited therein).

## 5.4 SIGNAL DIVERSITY RADAR

For practical reasons of implementation, radar systems typically transmit the same pulse-form over and over again. The standard ISAR approach assumes this kind of repetitious behavior and the development of sections 5.1 and 5.2 have further (implicitly) assumed a radar signal whose frequency representation is of the $S(\omega) \propto \mathrm{rect}(\omega - \omega_0)$ form of equation (2.7–8). One of the unfortunate consequences of this practice is the rather lengthy amount of time that the radar must 'dwell' on the target while waiting for it to rotate through the aspect angles that make up the synthetic aperture. In these systems, good cross-range resolution generally requires long data acquisition intervals and, for many applications, this is tactically unacceptable. Moreover, the target angular velocity $\Omega$ is generally not a measurable quantity and this means that the (ISAR) synthetic aperture is often unknown. Consequently, the cross-range dimension (the $x$-component of our ISAR image) is determinable only to within a scale factor. And despite the scattering-based exposition so far, two-dimensional radar imaging actually attempts to recover some target function $\varrho(T, \vartheta)$ (defined below) and $x$ and $y$ are related to the delay time $T$ and Doppler shift $\vartheta$ by $T = 2y/c$ and $\vartheta = 2k\Omega x$, respectively.

A scheme has been suggested [35, 36] which, in principle, could significantly reduce the dwell requirement of traditional ISAR by using *varying* pulse-forms as a basis set for estimating $\varrho(T, \vartheta)$. To see how this works, we need to redevelop the basic imaging equation (5.1–4) to explicitly display the role of the radar signal $s(t)$.

If $s(t)$ denotes an incident pulse transmitted by the radar, then the linear radar scattering model of equation (3.4–1) becomes

$$s_{\text{scatt}}(t) = \int_{\mathbb{R}^2} \varrho(t', \vartheta')\, s(t - t')e^{i\vartheta'(t-t')}\, dt'\, d\vartheta' \qquad (5.4\text{–}1)$$

where $\varrho(t, \vartheta)$ is the target reflectivity density defined in such a way that $\varrho(t, \vartheta)\, dt\, d\vartheta$ is proportional to the field reflected from the target at range between $ct$ and $c(t + dt)$ with Doppler shift bewteen $\vartheta$ and $\vartheta + d\vartheta$.

The radar matched filter response is easy to calculate and equation (2.5–8) becomes

$$
\eta(t) = \int_{\mathbb{R}} s_{\text{scatt}}(t')s^*(t'+t)\,dt'
$$

$$
= \int_{\mathbb{R}} \int_{\mathbb{R}^2} \varrho(t'',\vartheta'')\,s(t'-t'')e^{i\vartheta''(t'-t'')}s^*(t'+t)\,dt''\,d\vartheta''\,dt' \qquad (5.4\text{–}2)
$$

$$
= \int_{\mathbb{R}^2} \varrho(t'',\vartheta'')\chi(t+t'',\vartheta'')\,dt''\,d\vartheta''
$$

where $\chi$ was defined in equation (2.6–1). (Note that this is the generalized form of the radar mapping equation (2.5–8) associated with the scattering model of equation (5.4–1).)

Comparing equation (5.4–2) with the radar form of the 'shadow transform' (equation (5.1–2)), we observe that the two equations are formally the same when

$$
\chi(T,\vartheta) = (1/2\pi)\delta(\vartheta - \gamma T). \qquad (5.4\text{–}3)
$$

The projection line (denoted by $L$ in figure 5.1) in $x$–$y$ space is determined by the target aspect angle $\theta$. In $T$–$\vartheta$ space, the slope of this line is $\gamma$ and, substituting for $x$ and $y$ in terms of $T$ and $\vartheta$, we see that $\gamma$ and $\theta$ are related by the choice of equation (5.4–3) through

$$
\tan\theta = \frac{x}{y} = \frac{\vartheta/2k\Omega}{cT/2} = \frac{1}{\omega\Omega}\frac{\vartheta}{T} = \frac{\gamma}{\omega\Omega}. \qquad (5.4\text{–}4)
$$

Equations (5.4–2) and (5.1–2) are identical in situations where we can further claim that

$$
\varrho(T,\vartheta) = \overline{\rho}\left(\frac{\vartheta}{2k\Omega}, \frac{cT}{2}\right). \qquad (5.4\text{–}5)
$$

This last requirement will not be generally true, of course and, in particular, fails when $\varrho(T,\vartheta)$ contains scattering centers that do not persist over the whole of the synthetic aperture used to define the domain of $\overline{\rho}(x,y)$. It is often the case, however, that equation (5.4–5) is approximately true and one could even argue that determination of $\varrho(T,\vartheta)$ is more applicable to radar imaging applications than determination of $\overline{\rho}(x,y)$.

In [35, 36] it was recognized that the chirp waveform approximates the behavior of equation (5.4–3) with $\gamma$ the *chirp rate* (see section 2.7 and, in particular, equation (2.7–3)). Since chirp rate is a parameter controllable at the radar, the relationship between $\theta$ and $\gamma$ of equation (5.4–4) means that an effective 'synthetic aperture' can be created at the radar without the necessity of waiting for the target to rotate with time. Reconstruction of the target reflectivity density can be accomplished by using, for example, the convolution-backprojection algorithm (see section 5.1) or some other related technique. By rapidly transmitting a set of 'basis' pulses (with varying chirp rates), it may also be possible to significantly reduce the need for the target motion compensation that is normally required when the data are collected over much longer intervals.

While this formal connection between target aspect angle and radar chirp rate may, at first, appear bewildering, equation (5.4–2) and the subsequent analysis often helps to clarify the ideas behind ISAR imaging. Recall from our discussion of the ambiguity function that the uncertainty ellipse (section 2.6) expresses the fundamental inability of a single radar pulse to simultaneously achieve both arbitrarily high down-range and cross-range resolution. Small pulsewidth offers good down-range definition but poor cross-range feature isolation while small bandwidth offers the converse resolution properties. When the pulse-to-pulse shape is fixed, we can only get good down- and cross-range resolution by choosing either small pulsewidth or small bandwidth and sampling the target from different directions. Of course, when the pulse-to-pulse shape is allowed to vary so that we can, for example, transmit first a short-duration pulse (large bandwidth) and then a small-bandwidth pulse, we will be able to achieve good resolution in both dimensions (range and Doppler shift) without the need for the target to rotate—and pulse compression offers a convenient way to vary the transmitted signal's bandwidth. These two approaches can therefore be seen to be implementations of a much larger class of imaging schemes that we will return to in section 8.4.

# REFERENCES

[1]   Munson D C, O'Brien D and Jenkins W K 1983 A tomographic formulation of spotlight-mode synthetic aperture radar *Proc. IEEE* **71** 917
[2]   Merserau R M and Oppenheim A V 1974 Digital reconstruction of multi-dimensional signals from their projections *Proc. IEEE* **62** 1319
[3]   Brown W M and Porcello L J 1969 An introduction to synthetic aperture radar *IEEE Spectrum* **6** 52
[4]   Hovenessian S A 1980 *Introduction to Synthesis Array and Imaging Radar* (Dedham, MA: Artech House)
[5]   Mensa D, Heidbreder G, and Wade G 1980 Aperture synthesis by object rotation in coherent imaging *IEEE Trans. Nucl. Sci.* **27** 989
[6]   Chen C-C and Andrews H C 1980 Target-motion-induced radar imaging *IEEE Trans. Aerospace Electr. Syst.* **16** 2
[7]   Walker J L 1980 Range-Doppler imaging of rotating objects *IEEE Trans. Aerospace Electr. Syst.* **16** 23
[8]   Mensa D L 1981 *High Resolution Radar Imaging* (Dedham, MA: Artech House)
[9]   Wehner D R 1987 *High Resolution Radar* (Norwood, MA: Artech House)
[10]  Rihaczek A W and Hershkowitz S J 1996 *Radar Resolution and Complex-Image Analysis* (Dedham MA: Artech House)
[11]  Felsen L B 1987 Theoretical aspects of target classification *AGARD Lecture Series* no 152, ed L B Felsen (Munich: AGARD)
[12]  Langenberg K J, Brandfass M, Mayer K, Kreutter T, Brüll A, Fellinger P and Huo D 1993 Principles of microwave imaging and inverse scattering *EARSeL Adv. Remote Sensing* **2** 163

[13]    Steinberg B D 1988 Microwave imaging of aircraft *Proc. IEEE* **76** 1578
[14]    Cafforio C, Prati C and Rocca E 1991 SAR data focusing using seismic migration techniques *IEEE Trans. Aerospace Electr. Syst.* **27** 194
[15]    Brown W M and Fredricks R J 1969 Range-Doppler imaging with motion through resolution cells *IEEE Trans. Aerospace Electr. Syst.* **5** 98
[16]    Wehner D R, Prickett M J, Rock R G and Chen C C 1979 Stepped frequency radar target imagery, theoretical concept and preliminary results *Technical Report 490* (San Diego, CA: Naval Ocean Systems Command)
[17]    Prickett M J and Chen C C 1980 Principles of inverse synthetic aperture radar (ISAR) imaging *IEEE 1980 EASCON Record* (New York: IEEE) p 340
[18]    Chen C C and Andrews H C 1980 Multifrequency imaging of radar turntable data *IEEE Trans. Aerospace Electr. Syst.* **16** 15
[19]    Mensa D L, Halevy S and Wade G 1983 Coherent Doppler tomography for microwave imaging *Proc. IEEE* **71** 254
[20]    Ausherman D A, Kozma A, Walker J L, Jones H M and Poggio E C 1984 Developments in radar imaging *IEEE Trans. Aerospace Electr. Syst.* **20** 363
[21]    Kachelmyer A L 1992 Inverse synthetic aperture radar image processing *Laser Radar* **7** 193
[22]    Hua Y, Baqai F and Zhu Y 1993 Imaging of point scatterers from step-frequency ISAR data *IEEE Trans. Aerospace Electr. Syst.* **29** 195
[23]    Kong K K and Edwards J A 1995 Polar format blurring in ISAR imaging *Electron. Lett.* **31** 1502
[24]    Berizzi F and Corsini G 1996 Autofocusing of inverse synthetic aperture radar images using contrast optimization *IEEE Trans. Aerospace Electr. Syst.* **32** 1185
[25]    Kirk J C Jr 1975 Motion compensation for synthetic aperture radar *IEEE Trans. Aerospace Electr. Syst.* **11** 338
[26]    Bocker R P and Jones S 1992 ISAR motion compensation using the burst derivative measure as a focal quality indicator *Int. J. Imaging Syst. Tech.* **4** 286
[27]    Delisle G Y and Wu H 1994 Moving target imaging and trajectory computation using ISAR *IEEE Trans. Aerospace Electr. Syst.* **30** 887
[28]    Wu H, Grenier D and Fang D 1995 Translational motion compensation in ISAR processing *IEEE Trans. Image Proc.* **4** 1561
[29]    Wu H and Delisle G Y 1996 Precision tracking algorithms for ISAR imaging *IEEE Trans. Aerospace Electr. Syst.* **32** 243
[30]    Watanabe Y, Itoh T and Sueda H 1996 Motion compensation for ISAR via centroid tracking *IEEE Trans. Aerospace Electr. Syst.* **32** 1191
[31]    Trischman J A 1996 Real-time motion compensation algorithms for ISAR imaging of aircraft *SPIE Proc. on Radar/Ladar Proc. and Appl.* (Denver, CO)
[32]    Trischman J, Jones S, Bloomfield R, Nelson E and Dinger R 1994 An X-band linear frequency modulated radar for dynamic aircraft measurement *AMTA Proc.* (Long Beach, CA) (New York: AMTA) p 431
[33]    DeGraaf S R 1988 Parametric estimation of complex 2-d sinusoids *IEEE Fourth Ann. ASSP Workshop on Spectrum Estimation and Modeling* (New York: IEEE) p 391
[34]    Kleinman R E and van den Berg P M 1991 Iterative methods for solving integral equations *Radio Sci.* **26** 175
[35]    Bernfeld M 1984 Chirp Doppler radar *Proc. IEEE Lett.* **72** 540
[36]    Feig E and Grünbaum F A 1986 Tomographic methods in range-Doppler radar *Inverse Problems* **2** 185

# 6

# Model Errors and Their Effects

In chapter 5 we saw how polar reformatting and motion compensation can be used to correct for two common types of phase error in the measured ISAR data sets $\mathbf{H}$. These phase adjustments are required when the small-angle approximation is used, the target is large and/or the data are collected from targets undergoing translational motion—issues associated more with assumptions about the data than with assumptions about the scattering interactions between the incident field and the target. Mitigating these two sources of phase error across $\mathbf{H}$, followed by object function estimation based on equation (5.2–1), comprise the bulk of 'standard' ISAR image reconstruction processing.

Even if polar data and target motion were the only sources of error, however, practical application of ISAR imaging algorithms would not be straightforward. An unfortunate consequence of the coherent nature of radar waveforms is that target scattering centers can constructively and destructively interfere. This effect—known as 'scintillation'—often results in fluctuation and fading of range profile reference peaks and can complicate the range-bin alignment process. At high frequencies, typical radar data display a very sensitive dependence upon target aspect, and scintillation effects can dominate object function reconstruction. In addition, these scintillation effects also complicate ordinary HRR-based target identification methods since range profile variation means that *very many* 'template' profiles need to be stored in a target 'look-up' library against which measured data are to be compared.

Unlike the phase errors of chapter 5, scintillation effects can be thought of as 'second-order errors'—they are problems in target imaging and classification which usually manifest themselves when trying to correct for the 'first-order' consequences of a polar data domain or target motion. These second-order scintillation effects can be very important and must be carefully accounted for as part of any effective motion compensation scheme or template-based target recognition algorithm. In addition, the effects of scintillation are important in target tracking and associated target imaging methods, and we shall investigate the properties of scattering center interference in detail in chapter 7.

Scintillation induced errors can be understood within the context of the

71

standard weak scatterer model. This model will not always be accurate, of course, and another significant source of image error is associated with some of the approximations made in chapter 3 (in particular, that $\rho_{k,\hat{R}}$ can be treated as independent of $\theta$ and $k$). This approximation enables us to analyse equation (3.4–3) as a Fourier transform relation between $\rho$ and $H$ but results in large and difficult-to-interpret artifacts in both one- and two-dimensional imaging when the weak, isotropic scatterer approximation is invalid [1, 2].

## 6.1   TEMPLATE-BASED ATR

Before we can see how scintillation affects HRR-based target recognition, we must first understand how range profiles are applied to the problem of target identification and classification. Let $\tau_n(y; \theta)$ denote a 'template' representation of target '$n$' with orientation $\theta$. The template $\tau_n(y; \theta)$ can be calculated (in principle) from a detailed knowledge of the shape and scattering properties of target $n$, but it is more common to measure it in a carefully controlled laboratory environment. Template-based target classification methods employ a 'library' $\mathcal{L}_N$ of $N$ templates (one template for each potential 'target') against which new radar measurements are compared. We will assume that $\tau_n(y; \theta)$ is scaled with respect to all of the templates in the library so that each will have unit energy.

Let $\overline{\overline{\rho}}_{\text{est}}(y)$ be an HRR profile-based estimate of $\overline{\overline{\rho}}_\theta(y)$, which we also scale so that $\|\overline{\overline{\rho}}_{\text{est}}\| = 1$. Define the correlation coefficient $C(n, \theta)$ by

$$C(n, \theta) = \max_{y \in \mathbb{R}} \int_{\mathbb{R}} \overline{\overline{\rho}}_{\text{est}}(y') \tau_n(y' + y; \theta) \, dy' . \qquad (6.1\text{–}1)$$

By the Cauchy–Schwartz inequality, it follows that $C(n, \theta) \leqslant 1$ (equality can only occur when $\overline{\overline{\rho}}_{\text{est}}(y) = \tau_n(y; \theta)$ (up to a translation)). Equation (6.1–1) provides a tool for target identification, and we can assign

$$n = \arg_1 \max_{\substack{n' \in \mathcal{L}_N \\ \theta' \in (0, 2\pi)}} \left\{ C(n', \theta') \right\}$$

and $\qquad\qquad\qquad\qquad\qquad\qquad\qquad\qquad\qquad\qquad\qquad\qquad$ (6.1–2)

$$\theta = \arg_2 \max_{\substack{n' \in \mathcal{L}_N \\ \theta' \in (0, 2\pi)}} \left\{ C(n', \theta') \right\}$$

as the 'best guess' of what the target actually is. (Note that the relation (6.1–2) also yields an estimate of the target's orientation.)

The practical implementation of template-based target recognition is far more involved than represented by this simple discussion. For instance, equation (6.1–1) is usually implemented in a digital computer and the function $\overline{\overline{\rho}}_{\text{est}}(y)$ must be sampled on a grid $y \to y_i$. This may lead to registration problems when the grid is too coarse. In addition, measurement noise may corrupt the correlation coefficients to the point where equation (6.1–2) becomes unreliable. Usually, thresholds are set and the $C(n, \theta)$ are used to assign identification probabilities to the unknown target [3].

Because target orientation is not accurately estimable from HRR profiles by independent means, the library is required to contain not only different individual target representations for each possible target 'type', but also *different representations for each possible target orientation*. Moreover, while we have assumed the target to rotate only in the $x$–$y$ plane (for convenience of exposition), in reality aircraft targets can have any orientation $(\theta, \phi) \in (0, 2\pi) \times (0, \pi)$ and equations (6.1–1) and (6.1–2) must be modified to allow for this.

A library which must contain finely sampled templates to account for many targets and their possible orientations can quickly become very large. For example, if there are 25 possible aircraft target types, and orientations are sampled on a $10° \times 10°$ lattice, then the library will require about $25 \times 4\pi/(10\pi/180)^2 \sim 10^4$ stored representations $\tau_n(y; \theta)$. This is also the number of correlation coefficients $C(n, \theta)$ that must be calculated and, consequently, the storage and computation requirements are not inconsequential. In reality, the situation is often far worse.

## 6.2 UNRESOLVED SCATTERERS AND SCINTILLATION

Implicit in the template-based HRR approach to target identification is the assumption that the range profiles vary smoothly as the target aspect changes (so that, for example, $\tau_n(y; \theta)$ need only represent the target at $10°$ intervals). That this is not true (often, spectacularly not true) is not a consequence of any model error resulting from the approximations that were made in chapter 3. Rather, rapid variation of the radar return (scintillation) can be understood within the framework of weak isotropic scatterers.

When two or more scatterers lie within the same resolution cell (intuitively defined by the pulsewidth on the target), they will appear to be only one scattering center in a range profile—this is, after all, what range-resolution means. The scattered field, however, will not be that associated with a single scatterer and will instead be the sum of two (or more) complex-valued response pulses which can add constructively or destructively. In the two-scatterer case, for example, we can use equations (3.2–7) and (3.4–3) to write (small-angle approximation)

$$
\begin{aligned}
H(k, \theta) &\approx A_1 e^{-i2kx_1\theta} + A_2 e^{-i2kx_2\theta} \\
&= e^{-ik\theta(x_1+x_2)} \left( A_1 e^{ik\theta(x_2-x_1)} + A_2 e^{-ik\theta(x_2-x_1)} \right) \\
&= e^{-ik\theta(x_1+x_2)} \Big[ 2A_1 \cos[k\theta(x_2 - x_1)] \\
&\qquad\qquad + (A_2 - A_1)e^{-ik\theta(x_2-x_1)} \Big].
\end{aligned}
\tag{6.2–1}
$$

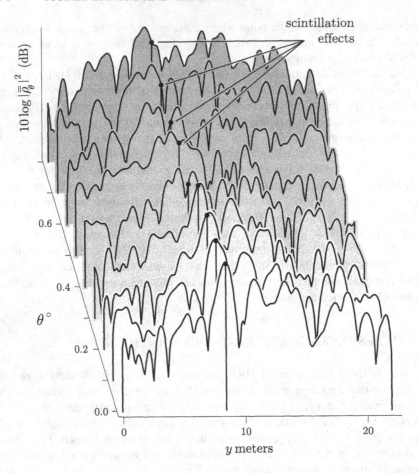

**Figure 6.1**   Scintillation in range profiles as a function of aspect. A peak that is clearly defined at aspects between $0°$ and $0.3°$ (marked in the figure) can be seen to change in magnitude and position for aspects greater than about $0.4°$.

When $A_1 \sim A_2$, equation (6.2–1) yields

$$|H(k, \theta)|^2 \approx |2A_1|^2 \cos^2[k\theta(x_2 - x_1)] \qquad (6.2\text{–}2)$$

and, consequently, even very small changes in $\theta$ can result in dramatic variation in $|H(k, \theta)|$ when the product $k(x_2 - x_1)$ is large. (The situation for more than two scatterers, or when $A_1 \not\approx A_2$, can be more complex but the main idea is unchanged. We shall return to this in chapter 7 where we will show that the behavior described by equation (6.2–2) is generally true and $H$ is a sine-type function.)

Figure 6.1 illustrates scintillation-induced range profile changes as the aspect is varied through about $1°$. The data are the same as those used in figures 5.5 and 5.6 (the B-727 jetliner with $\overline{\omega} = 2\pi \times 9.25$ GHz and $\Delta\omega = 2\pi \times 500$ MHz). The most obvious of the scintillation effects are manifest as amplitude variations associated with the largest peak (these are marked in the figure).

The true problem of HRR profile-based target identification can now be appreciated. The library templates $\tau_n(y; \theta)$ may need to be created at as many aspects as a profile $\overline{\overline{\rho}}_\theta(y)$ can be expected to display 'significant' variation. And, because of scintillation effects, $\overline{\overline{\rho}}_{est}(y)$ can have a very sensitive dependence on target aspect. We can use equation (6.2–2) to get an estimate of how dense such a library aspect grid will need to be. If we assume that the main scattering element of $\overline{\overline{\rho}}_{est}$ is due to two equal-amplitude point scattering centers separated in their cross-range dimension by 1 m and lying within the same range bin, and our correlation-based identification algorithm can tolerate variations in this peak of as much as 50% ($|H(k, \delta\theta)|/|H(k, 0)| = 0.5$), then $\delta\theta < \cos^{-1}(0.5)/k$. At typical radar frequencies ($2\pi \times 10^{10}$ Hz), this implies that $\delta\theta < 0.5°$. Practical experience confirms this estimate and requires our example 25-target type template library to now store $> 10^6$ target representations $\tau_n(y; \theta)$.

The template 'storage problem' is usually addressed by using 'reduced templates' to represent the known targets. Ideally, these reduced templates will be 'minimal' in the sense that they will encode only the information about the target that is required for identification and not the entire profile itself. For example, the templates might be reduced to the value and location of the five strongest scattering centers in a profile and, consequently, each would be expressible as ten numbers (five positions + five amplitudes). (The CLEAN algorithm automatically prioritizes range profile peaks in this way.) Of course, the correlation coefficients of equation (6.1–1) must still be computed for each template and this is often the true bottleneck for real-time applications. Further speed improvement typically involves library-size reduction.

One often-proposed solution to this problem is HRR 'profile-averaging' which assigns representative templates on a coarse aspect grid based on an average of templates on a fine grid. Because of the behavior described by equation (6.2–2) (see also equation (7.3–2)), scintillation is a cyclostationary process [4]. Consequently, the *statistics* of multiple range profiles, collected from a narrow range of target aspects, can be expected to vary slowly throughout a suitably chosen library. (The details of this dependance will still generally depend on large target aspect variations and the behavior is really 'quasi-stationary'.) The usual approach is to employ variable grid spacing so that template grid density will be determined by their rate of variation as functions of aspect. The general idea behind these ideas has come to be known as 'clustering' [5].

In ISAR imaging it is often possible to resolve scattering centers in both range and cross-range; scintillation-induced amplitude fading in high-resolution two-dimensional images is not nearly as common as in one-dimensional images.

Scintillation *is* a problem for the motion-compensation component of ISAR reconstruction algorithms, however, and can seriously affect phase correction across **H**. This is because range profile alignment can be thought of as assigning shifts to profile peaks (recall equation (5.4–1) and the subsequent discussion) and requires these peaks to be well defined over the synthetic aperture. Scintillation of these peaks can adversely affect range alignment which means that the motion-induced phase errors may go uncorrected. In ISAR imaging, scintillation effects can result in image distortion and artifacts.

Scintillation is considered to be a 'standard' problem in coherent imaging and its consequences must be addressed as a fundamental limitation of radar measurement systems—they are manifestly independent of the target model used. Other errors turn out to be a result of attempting to interpret the resultant images within the context of the weak scatterer approximation.

## 6.3  NON-WEAK AND DISPERSIVE SCATTERERS

The weak, isotropic, frequency-independent scatterer approximation of equation (5.1–3) is often sufficient when the imaging requirements are not too demanding. But many contemporary target identification schemes require a high level of system resolution and sensitivity and the limitations of this analytically 'convenient' model can adversely affect the target recognition process by introducing image artifacts that are displaced from the target component responsible for them. It is a simple fact that many target subcomponents are not weak point scatterers and, generally, $\rho = \rho(x; k, \theta)$.

That the shape of $\rho$ generally displays a dependence on $\theta$ is clearly evident from an inspection of equation (3.2–6) in which $\hat{R} \cdot x_{\mathrm{sp};m}$ will depend on target aspect for all but the simplest bodies. Moreover, target substructure elements will sometimes scatter strongly in only one direction and may be observed to 'turn on' and 'turn off' as the target rotates (recall the discussion following equation (3.1–11)). In addition, if the 'weak scatterer' approximation is not valid, then significant multiple scattering events can occur resulting in time delays which will appear to displace the image element in range from its associated target subscatterer. (Multiple scattering-induced artifacts are illustrated by the idealization of figure 6.2.)

Perhaps the best example of multiple scattering induced artifacts is associated with 'duct dispersion' [5–16]. In general, 'dispersion' refers to wave propagation in media where the wave speed depends upon the wave frequency. Since the speed at which a wave travels down a waveguide has such a dependence, the field scattered from a target component within a re-entrant structure (duct or cavity) will return to the radar at a different (delayed) time from the scattered field components due to the rest of the target. In both HRR and ISAR imaging, duct dispersion artifacts appear as multiple time-delayed image elements which often extend beyond the (down-range) support of the target. Since the portion of

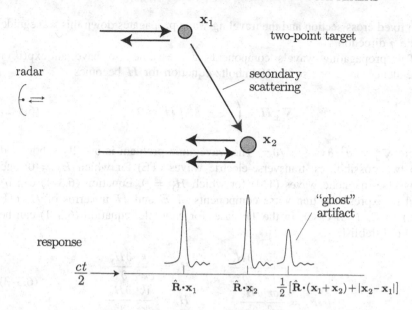

two-point target

radar

secondary scattering

"ghost" artifact

response

$$\frac{ct}{2}$$

$\hat{\mathbf{R}}\cdot\mathbf{x}_1$      $\hat{\mathbf{R}}\cdot\mathbf{x}_2$      $\frac{1}{2}[\hat{\mathbf{R}}\cdot(\mathbf{x}_1+\mathbf{x}_2)+|\mathbf{x}_2-\mathbf{x}_1|]$

**Figure 6.2** Strong multiple scattering terms can introduce artifacts that, under the weak scatterer assumption, will not be correctly mapped to target structure elements. This is illustrated by the figure in which secondary scattering (from scatterer 1 off of scatterer 2) is mistakenly assigned to a spurious third scatterer.

the radar field that finds its way into the entrance of a target duct or cavity will be be unable to scatter isotropically—and, actually, may scatter preferentially back in the direction of the radar—the artifacts associated with these structures can be the strongest features in the radar image and may even obscure other image elements. Note that the example images of figures 4.1 and 5.6 display some of these image artifacts associated with dispersive scattering from the jet engines.

A common model for duct dispersion relies on a 'ray tracing' interpretation of scattering and attributes the time delays of the field scattered from re-entrant structures to extended path lengths of rays which 'bounce around' the interior of the duct or cavity. More traditional analysis applies the Maxwell equations

$$\nabla \times E = \frac{i\omega}{c}H$$
$$\nabla \times H = -\frac{i\omega}{c}E$$

(6.3–1)

(where we have assumed an $\exp(-i\omega t)$ time dependence). We will illustrate duct dispersion by considering a simple example in which the duct is a waveguide with fixed cross-section and the traveling wave propagates down this waveguide in the $\hat{\jmath}$ direction.

The propagating wave's components are assumed to have an $\exp(iky)$ dependence on $y$, and so the Helmholtz equation for $H$ becomes

$$\nabla_\perp^2 H + \left( \frac{\omega^2}{c^2} - k^2 \right) H = 0 \tag{6.3–2}$$

where $\nabla_\perp^2 = \partial^2/\partial x^2 + \partial^2/\partial z^2$. The waveguide problem is usually subdivided into two possibilities: transverse-electric waves (TE) for which $E_y = 0$; and transverse-magnetic waves (TM) for which $H_y = 0$. Equation (6.3–1) can be used to express the transverse components of $E$ and $H$ in terms of $H_y$ (TE case) or $E_y$ (TM case). In the TE case, for example, equation (6.3–1) can be used to establish

$$E_x = -\frac{i\omega}{c\kappa^2} \frac{\partial H_y}{\partial z}, \qquad H_x = \frac{ik}{c\kappa^2} \frac{\partial H_y}{\partial x}$$
$$E_z = \frac{i\omega}{c\kappa^2} \frac{\partial H_y}{\partial x}, \qquad H_z = \frac{ik}{c\kappa^2} \frac{\partial H_y}{\partial z} \tag{6.3–3}$$

where

$$\kappa^2 = (\omega/c)^2 - k^2. \tag{6.3–4}$$

Consequently, all of the field components are determined from the solution to

$$\nabla_\perp^2 H_y + \kappa^2 H_y = 0 \tag{6.3–5}$$

subject to the boundary conditions of equation (3.1–1).

The simplest example is a waveguide consisting of two perfectly conducting infinite parallel plates separated by a distance $d$. We choose a geometry in which the $y$–$z$ plane is parallel to the plates which lie at $x = 0$ and $x = d$. In the TE case, the perfect conductor boundary condition on $H_y$ can be deduced from equation (6.3–3): no normal components of $H$ at the boundary means that $H_x|_{x=0,d} = 0$, and so (using equation (6.3–3))

$$\frac{\partial H_y}{\partial x}\Bigg|_{x=0,d} = 0. \tag{6.3–6}$$

Equations (6.3–5) and (6.3–6) specify an eigenvalue problem. The solutions are

$$H_y(x, y, z) = H_0 e^{i(ky-\omega t)} \cos \kappa x \tag{6.3–7}$$

with the eigenvalues

$$\kappa_m^2 = \left( \frac{m\pi}{d} \right)^2, \quad m = 1, 2, \dots. \tag{6.3–8}$$

The velocity of wave propagation is given by $v = \partial\omega/\partial k$. Substituting equation (6.3–8) into equation (6.3–4) and differentiating yields

$$v_m = \frac{\partial\omega}{\partial k} = \frac{ck}{\sqrt{\left(\frac{m\pi}{d}\right)^2 + k^2}}, \quad m = 1, 2, \dots \quad (6.3\text{–}9)$$

From this result we can see that, as $m$ increases, the wave speed will decrease and scattering centers lying between the two plates will contribute terms to $H_{scatt}$ that are time-delayed with respect to contributions from (equirange) scattering centers that lie outside the plates.

The eigenfunctions of equation (6.3–6) are called the 'modes' of the waveguide and are indexed by $m$. For a given frequency $\omega$, the wave number $k$ is given by

$$k_m^2 = \left(\frac{\omega}{c}\right)^2 - \kappa_m^2. \quad (6.3\text{–}10)$$

If $\omega < c\kappa_m$, then $k_m$ will be imaginary and there will be no propagating wave. For this reason, $\omega_m \equiv c\kappa_m$ is called the 'cut-off' frequency.

In general, the cross-sectional geometry of the duct or cavity will define the mode behavior and, consequently, the speed of propagation within the re-entrant target structure. Since each mode will support waves that travel at different speeds, a target's duct(s) or cavity(ies) can contribute multiple components to the estimated image—each component time delayed (i.e. range shifted) with respect to the others. When the cross-sectional geometry is not constant along the length of the duct, then the simple analysis that we applied in equations (6.3–5)–(6.3–9) will, generally, no longer apply—but dispersive wave behavior will still occur. (If the variation in cross section is slow, however, it may still be possible to describe re-entrant propagation in terms of 'quasi-modes'.)

## 6.4  CORRECTIVE PSF

The HRR and ISAR scattering equations (4.1–2) and (5.1–3) only apply to the problems of imaging weak, isotropic and frequency-independent scatterers—indeed, for the case of frequency-dependent $\rho$, all of the results of chapters 4 and 5 are generally inapplicable. Because of its ease of interpretation, however, it is sometimes advantageous to save as much of the 'fixed-point scatterer' nature of $\rho$ as we can. An approach to accommodating the target's general subscatterer effects fixes an effective 'phase center' to a single location on the target and describes possible variations in $\rho$ by including a dispersive phase term in the radar waveform. In this way, we may consider an object function $\rho(x) = \rho(x; k_0, \theta_0)$ to be independent of $k$ and $\theta$, and account for the position shift over $\theta$ and $k$ by a general dispersion relation $k\hat{R} \cdot x \rightarrow \beta(x, \theta, k)$. Substituting the plane wave interrogating waveform into equation (3.2–2) and applying the same arguments that led to equation (5.2–3) now yields

$$H(k,\theta) \approx \frac{1}{(2\pi)^2} \int_{\mathbb{R}^2} \int_{\mathbb{R}^2} \overline{\rho}\left(x',y'\right) W(x',y',x'',y'')$$
$$\times e^{-i(k_x x'' + k_y y'')}\, dx'\, dy'\, dx''\, dy'' \tag{6.4-1}$$

where

$$W(x',y',x'',y'') = \int_{\mathbb{K}} S(x',\theta',k') e^{i(k_x' x' + k_y' y' - 2\beta(x',\theta',k'))}$$
$$\times e^{i(k_x'(x''-x') + k_y'(y''-y'))}\, dk_x'\, dk_y' \tag{6.4-2}$$

and $S(x,\theta,k) \leqslant 1$ accounts for varying strength (i.e. 'modal strength'). Equation (6.4–1) should be compared with equation (5.2–3) and demonstrates how the effects of complex (frequency-dependent) scattering centers can be interpreted as being due to a spatially varying point-spread function $W(x,x')$ acting on an ideal object function consisting of weak, isotropic and *frequency-independent* scatterers fixed at effective phase-center locations on the target.

Alternately, it is sometimes held that the 'effective' target object function $\rho_{\text{eff}} = \rho$ (obtained from equation (3.4–3) without regard to the model limitations) is all that is necessary for practical target identification. As can be seen from figure 5.6, $\rho_{\text{eff}}$ will generally have a support that extends beyond the physical boundaries of the actual target and may bear little resemblance to the target itself. This argument contends that automatic (i.e. machine-based) target recognition schemes will compare $\rho_{\text{eff}}$ to a library of known object functions to establish a match and that this library could just as well consist of 'effective' object functions. Of course, this point of view is potentially useful but is really only relevant *after the range walk problem has been addressed*. Since range-bin alignment requires that the reference peak of each aspect-dependent range profile be fixed to the location of a unique target feature, this initial data correction processing requires that the point-like scattering centers be identified and separated from the non point-like scatterers. Moreover, in the simpler (1D) images (that do not require range-bin alignment) the strength of these multiple duct features can obscure important target identification features and reliable classification methods may need to restrict the *usable* image features to those target elements which appear in 'front' of the duct entrance.

## 6.5  DUCTS AND CAVITIES

Various signal processing techniques have been proposed to address the problems caused by duct dispersion [17–20]. As an example of how equation (6.4–1) can be used to account for image artifacts, we will examine scattering from target ducts and cavities (this was discussed in section 6.3) [21]. The point-spread function of equation (6.4–2) associated with terminated waveguides of 'length' $L$, with entrance located at $x_{\text{inlet}}$ (which we shall take as the phase center associated with this structure), can be expressed in terms of a modal expansion: a sum over the eigenfunctions of the waveguide. If $x = x_{\text{inlet}}$, then we can replace

$$S(x, \theta, k)e^{i(k_x x + k_y y - 2\beta(x,\theta,k))}$$

by

$$\frac{1}{k} \sum_{m=1}^{M} S_m(\theta, k)e^{i(k_x x + k_y y - 2\beta_m(x,\theta,k))}.$$

(6.5–1)

The phase term $\beta_m$ depends upon the cut-off frequencies $\omega_m = c\kappa_m$ which are determined by the cross-sectional geometry of the waveguide and are labeled by the mode indices $m$. The cut-off frequencies also allow us to limit the number of terms in the modal sum to $M = \max\{m | \kappa_m \leqslant k_2\}$, where $k_2$ is the largest $k$ in $\mathbb{K}$. The non-dispersive scatterer assumption incorrectly sets $\omega = ck_y$ for the image reconstruction. Equation (6.3–10) allows us to correct for this by setting

$$2\beta_m(x, \theta, k) = \begin{cases} k_x x - L\sqrt{k_y^2 - (2\kappa_m)^2}, & \text{if } x = x_{\text{inlet}} \\ \\ k_x x + k_y y, & \text{otherwise.} \end{cases}$$

(6.5–2)

This approximation neglects multiple scattering between the open and closed ends of the waveguide but is otherwise known to be quite good [13]. The strength factor $S_m$ is proportional to the energy which couples into the waveguide and generally displays a projected-area variation with $\theta$. The dependence of $S_m$ on $k$ is typically more complex than $\beta_m$ but also has a smaller effect than the phase factor of equation (6.5–1). If we assume that $S_m$ is (approximately) independent of $k$, then we can multiply equation (6.4–1) by $e^{i(k_x x + k_y y)}$ and integrate over $\mathbb{K}$ to obtain

$$\hat{\rho}(x, y) \approx \int_{\mathbb{R}^2} dx' dy' \, \overline{\rho}(x', y') \, \text{sinc}[\overline{k}\Delta\theta(x - x')]$$

$$\times \begin{cases} \sum_{m=1}^{M} S_m \int_{2k_1}^{2k_2} k_y^{-1} e^{iL\sqrt{k_y^2 - (2\kappa_m)^2}} e^{ik_y y} \, dk_y & \text{if } x' = x_{\text{inlet}} \\ \\ \text{sinc}[\Delta k(y - y')]e^{i2\overline{k}(y - y')} & \text{otherwise} \end{cases}$$

(6.5–3)

where $\overline{k} = (k_1 + k_2)/2$, $\Delta k = k_2 - k_1$, and $\Delta\theta = \theta_2 - \theta_1$ are defined by the boundary of $\mathbb{K}$ (which is assumed to be rectangular).

The image $\hat{\rho}(x, y)$ can be understood in terms of the object $\overline{\rho}(x_{\text{inlet}}, y)$ by examining the integral

$$I_m(y) = \int_{2k_1}^{2k_2} k_y'^{-1} e^{iL\sqrt{k_y'^2 - (2\kappa_m)^2}} e^{ik_y' y} \, dk_y'.$$

(6.5–4)

If we substitute [22]

$$e^{iL\sqrt{k_y^2-(2\kappa_m)^2}} = i\sqrt{k_y^2 - (2\kappa_m)^2}$$
$$\times \int_L^\infty J_0\left(2\kappa_m\sqrt{y''^2 - L^2}\right)e^{-ik_y y''}\, dy'' \tag{6.5-5}$$

where $J_0(\xi)$ is the Bessel function of the first kind, then we can write

$$I_m(y) = i \int_L^\infty dy''\, J_0\left(2\kappa_m\sqrt{y''^2 - L^2}\right)$$
$$\times \int_{2\max(k_1,\kappa_m)}^{2k_2} dk_y'\,\sqrt{1 - (2\kappa_m/k_y')^2}e^{ik_y'(y-y'')}. \tag{6.5-6}$$

(Wave components below cut-off will be exponentially damped and the lower bound of equation (6.5–6) has been altered so that only the components with $k_y > 2\kappa_m$ are included.)

Equation (6.5–6) expresses $I_m(y)$ as a convolution of $J_0(2\kappa_m\sqrt{y^2 - L^2})$ with $\int_{2\max(k_1,\kappa_m)}^{2k_2}\sqrt{1-(2\kappa_m/k_y)^2}e^{ik_y y}\, dk_y$ and is somewhat easier to interpret than equation (6.5–4). If we make the approximation

$$\int_{2\max(k_1,\kappa_m)}^{2k_2}\sqrt{1-(2\kappa_m/k_y')^2}e^{ik_y'(y-y'')}\, dk_y'$$
$$\approx \alpha\Delta k_{(1,c)}\,\text{sinc}[\Delta k_{(1,c)}(y-y'')]e^{i2\overline{k}_{(1,c)}(y-y'')} \tag{6.5-7}$$

where $\overline{k}_{(1,c)} = (\max(k_1,\kappa_m) + k_2)/2$ and $\Delta k_{(1,c)} = k_2 - \max(k_1,\kappa_m)$, then it is easy to see that the down-range effects of the inlet will appear as the function $J_0(y)$, shifted according to $y \to \sqrt{y^2 - L^2}$, dilated by $y \to 2\kappa_m y$, and 'blurred' by the $\text{sinc}(\Delta k_{(1,c)}y)$ function.

Figure 6.3 graphically displays the accuracy of the approximation of equation (6.5–7). Note that when $\kappa_m \ll k_1$ the approximation is very good. Moreover, when $\kappa_m \sim k_1$, the general effect is to scale the $\text{sinc}(\Delta k_{(1,c)}y)$ function by a factor $\alpha < 1$ while closely retaining its general shape. ($\alpha$ can be estimated by expanding the radical in the integrand of equation (6.5–7) but we will not require this result here.) Since $\kappa_m$ typically increases with increasing $m$, this means that higher modes will contribute proportionally less to the image $\hat{\rho}(x, y)$ and that modes for which $\kappa_m \geq k_1$ will be reduced by $\alpha \approx 0.5$. In addition, the scale factor $\Delta k_{(1,c)} = k_2 - \max(k_1,\kappa_m)$ means that when $\kappa_m > k_1$ the $\text{sinc}(\Delta k_{(1,c)}y)$ blurring function will cause the associated image elements to be less sharply defined.

Because inlet-induced artifacts are not point-like—they appear as extended image structures—they are sometimes moderated by filtering in the (Fourier) frequency domain. Such filtering relies on the assumption that all other scatterers

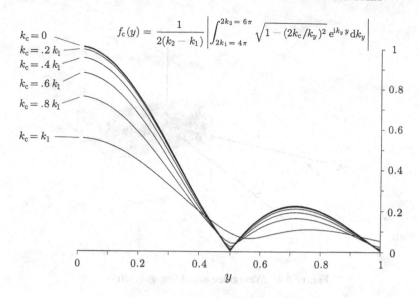

$$f_c(y) = \frac{1}{2(k_2 - k_1)} \left| \int_{2k_1 = 4\pi}^{2k_2 = 6\pi} \sqrt{1 - (2k_c/k_y)^2} \, e^{ik_y y} dk_y \right|$$

$k_c = 0$
$k_c = .2\,k_1$
$k_c = .4\,k_1$
$k_c = .6\,k_1$
$k_c = .8\,k_1$
$k_c = k_1$

**Figure 6.3** The sinc($y$) approximation of equation (6.5–7).

*are* point-like so any frequency domain 'peaks' must be due to the dispersive target elements. The results of equations (6.5–6) and (6.5–7) show that such filtering may not be reliable, however, and frequency domain searches can miss important modes which may be poorly distinguished because of the underlying Bessel function model. Rather, our analysis implies that we should concentrate on estimating $L$ and $\kappa_m$ within the framework of this Bessel function model. And although there exists an orthogonality relation for $J_0(\xi)$, it cannot be applied (in a straightforward manner) toward estimating $L$ and $\kappa_m$ because of the form of the argument of $J_0(2\kappa_m\sqrt{y^2 - L^2})$. The $Y$-transform, however, can be employed to show that [22]

$$-\pi\kappa \int_L^\infty J_0\left(2\kappa_m\sqrt{y'^2 - L^2}\right) y'^n Y_1\left(2\kappa y'\right) dy'$$

$$= \begin{cases} K_0\left(L\sqrt{(2\kappa_m)^2 - (2\kappa)^2}\right) & \text{if } \kappa_m > \kappa \\[2ex] -\frac{\pi}{2}(-1)^n Y_0\left(L\sqrt{(2\kappa)^2 - (2\kappa_m)^2}\right) & \text{if } \kappa > \kappa_m \end{cases} \tag{6.5–8}$$

where $Y_\nu(\xi)$ is the Bessel function of the second kind (of order $\nu$), $K_0(\xi)$ is the modified Bessel function of order 0, and $n = 0, 1, \dots, \infty$. Since $K_0(\xi)$ and $-Y_0(\xi)$ both rise sharply to $\infty$ as $\xi \to 0$, this result suggests

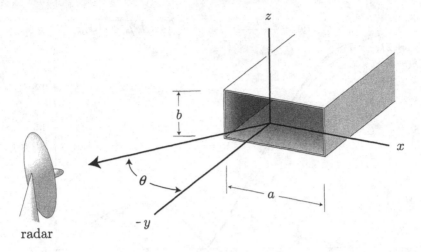

**Figure 6.4**  Waveguide scattering geometry.

a method for extracting the parameters $\kappa_m$ from the image $\hat{\rho}(x_{\text{inlet}}, y) \sim \sum_m s_L(y) J_0(2\kappa_m\sqrt{y^2 - L^2})$ where $s_L(y)$ denotes the unit step at $y = L$.

Define the transformation

$$\mathcal{P}_n(\kappa) = -\pi\kappa \int_0^\infty \hat{\rho}(x_{\text{inlet}}, y')y'^m Y_1(2\kappa y')\, dy' \qquad (6.5\text{--}9)$$

for $\kappa > 0$. Then it is easy to see from equations (6.5–3)–(6.5–8) that $\mathcal{P}_n(\kappa)$ will be sharply peaked when $\kappa = \kappa_m$, $k_1 \leqslant \kappa \leqslant k_2$.

The *reciprocal* transform to equation (6.5–9) is not quite as well defined, however. Denote the Struve function by $\boldsymbol{H}_\nu(\cdot)$. It can be shown [23] that

$$\mathcal{H}_\nu\{f\}(y) = \int_0^\infty f(\kappa')\boldsymbol{H}_\nu(y\kappa')(y\kappa')^{1/2}\, d\kappa' \qquad (6.5\text{--}10)$$

obeys $\mathcal{H}_\nu\mathcal{Y}_\nu = \mathcal{Y}_\nu\mathcal{H}_\nu = \mathcal{I}$ for functions in $L^2(0, \infty)$ when $-1 < \nu < 0$.

Of course, our obvious goal is to develop some means of removing duct artifacts while leaving non-dispersive scatterers unaltered. Equations (6.5–9) and (6.5–10) suggest a possible $\kappa$-domain filtering approach and these results were tested on data collected from a waveguide scattering experiment. We launched a (stepped-frequency) pulse at the open end of a terminated 103 cm length of rectangular waveguide (see figure 6.4). This waveguide has dimensions $a = 7.2\,\text{cm}$ and $b = 3.4\,\text{cm}$ and, consequently, has cut-off frequencies defined by [24]

$$\kappa_{m,n} = \sqrt{\left(\frac{m\pi}{a}\right)^2 + \left(\frac{n\pi}{b}\right)^2} = \sqrt{0.1890\, m^2 + 0.8454\, n^2} \qquad (6.5\text{--}11)$$

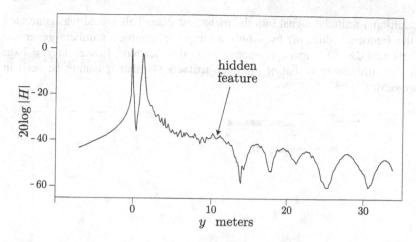

**Figure 6.5**  Measured data.

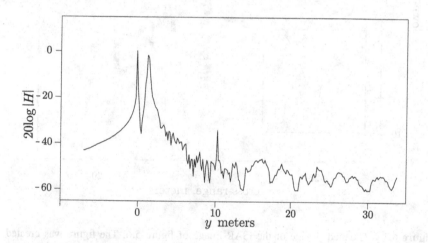

**Figure 6.6**  'Cleaned' image of a dispersive scatterer.

(in cm$^{-1}$) where $m, n = 1, 2, \ldots$ for TM modes and $m, n = 0, 1, 2, \ldots, (m, n$ not both 0), for TE modes.

Figure 6.5 shows the echo returned from the waveguide when the frequency band was $\Delta\omega = 2\pi \times (12.16, 13.26)$ GHz and aspect angle $\theta \approx \pi/6$. Plotted is the magnitude of the Fourier transform of the scattered field data (in dB with range increasing to the right). The duct-related artifacts can be seen extending for many multiples of the length of the target. For test purposes we introduced an additional artificial signal into the measured data. This signal was generated (in the frequency domain) by simply adding the complex numbers generated by $0.02 \exp(i2k_j * 10)$ and corresponds to a point scatterer located at $y = 10\,\text{m}$ with strength less than that of the duct artifacts (so that it cannot be seen in figure 6.5).

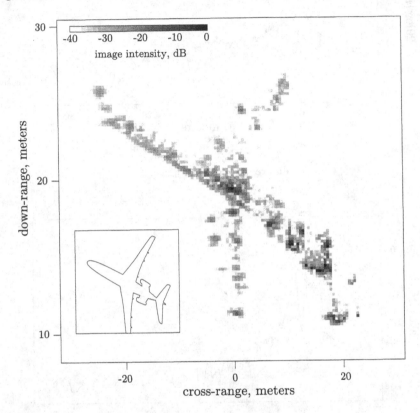

**Figure 6.7** 'Cleaned' image of the ISAR image of figure 5.6. The figure was created from the data used to generate figure 5.6 by applying a 'smooth' truncation filter (see text) to the $\mathcal{P}_2$-transform of each of the down-range cuts in the right half of the image and then applying the $\mathcal{H}_2$-transform to return to the image domain.

The image artifacts in figure 6.5 are to be reduced by filtering in the $\mathcal{P}_n$-transform domain. As part of this example, we applied a simple truncation filter: $\mathcal{P}_2(\kappa) \rightarrow 0$ when $|\mathcal{P}_2(\kappa)| > \tau$ where $\tau$ is a user defined threshold. Figure 6.6 is the 'cleaned' reconstruction formed by inverting the filtered results with $\tau = 0.3 \times \max\{|\mathcal{P}_2(\kappa)|\}_\kappa$. Note that the point scatterer at $y = 10\,\text{m}$ is now clearly visible and that the duct-related artifacts have indeed been reduced.

A potentially more useful example will be one which applies the mitigation filter to an actual ISAR image. For this purpose, we choose the data that were used to generate the ISAR image example of figure 5.6 (the B-727 jetliner) and which display artifacts associated with the jet engines. The peak value $\mathcal{P}_{max} = \max_{\kappa} \mathcal{P}_n(\kappa)$ was first determined and the smooth truncation filter $\mathcal{P}_n(\kappa) \to \mathcal{P}_n(\kappa) \times [(\mathcal{P}_{max} - |\mathcal{P}_n(\kappa)|)/\mathcal{P}_{max}]^8$ was applied to each of the down-range 'cuts' associated with the right half of figure 5.6 with $n = 2$. The (Struve) inverse transform was used to regain the (now filtered) image and the overall result appears in figure 6.7. Comparison of figures 5.6 and 6.7 demonstrates a significant artifact reduction.

# REFERENCES

[1] Jain A and Patel I 1990 Simulations of ISAR image errors *IEEE Trans. Instrumentation Meas.* **39** 212
[2] Dural G and Moffatt D L 1994 ISAR imaging to identify basic scattering mechanisms *IEEE Trans. Antennas Propag.* **42** 99
[3] Hudson S and Psaltis D 1993 Correlation filters for aircraft identification from radar range profiles *IEEE Trans. Aerospace Electr. Syst.* **29** 741
[4] Gardner W 1990 *Introduction to Random Processes* 2nd edn (New York: McGraw-Hill)
[5] Pham D T 1998 Applications of unsupervised clustering algorithms to aircraft identification using high range resolution radar *Non-Cooperative Air Target Identification Using Radar, AGARD Symp. (22–24 April, Mannheim, Germany)* (Quebec: Canada Communication Group Inc) p 16
[6] Witt H R and Price E L 1968 Scattering from hollow conducting cylinders *Proc. Inst. Electr. Eng.* **115** 94
[7] Moll J W and Seecamp R G 1969 Calculation of radar reflecting properties of jet engine intakes using a waveguide model *IEEE Trans. Aerospace Electr. Syst.* **6** 675
[8] Johnson T W and Moffatt D L 1982 Electromagnetic scattering by open circular waveguides *Radio Sci.* **17** 1547
[9] Huang C-C 1983 Simple formula for the RCS of a finite hollow circular cylinder *Electron. Lett.* **19** 854
[10] Pathak P H, Chuang C W and Liang M C 1986 Inlet modeling studies *Ohio State University ElectroScience Lab Report 717674-1 October 1986*
[11] Ling H, Chou R-C and Lee S W 1989 Shooting and bouncing rays: calculating the RCS of an arbitrary cavity *IEEE Trans. Antennas Propag.* **37** 194
[12] Wang T M and Ling H 1991 Electromagnetic scattering from three-dimensional cavities via a connection scheme *IEEE Trans. Antennas Propag.* **39** 1501
[13] Pathak P H and Burkholder R J 1991 High-frequency electromagnetic scattering by open-ended waveguide cavities *Radio Sci.* **26** 211
[14] Ross D C, Volakis J L and Hristos T A 1995 Hybrid finite element analysis of jet engine inlet scattering *IEEE Trans. Antennas Propag.* **43** 277
[15] Moore J and Ling H 1995 Super-resolved time-frequency analysis of wideband backscattered data *IEEE Trans. Antennas Propag.* **43** 623
[16] Kim H and Ling H 1993 Wavelet analysis of radar echo from finite-size targets *IEEE Trans. Antennas Propag.* **41** 200

[17]  Moghaddar A and Walton E K 1993 Time-frequency distribution analysis of scattering from waveguides *IEEE Trans. Antennas Propag.* **41** 677

[18]  Carin L, Felsen L B, Kralj D, Pillai S U and Lee W C 1994 Dispersive modes in the time domain: analysis and time-frequency representation *IEEE Microwave Guided Wave Lett.* **4** 23

[19]  Trintinalia L C and Ling H 1996 Extraction of waveguide scattering features using joint time-frequency ISAR *IEEE Microwave Guided Wave Lett.* **6** 10

[20]  Trintinalia L C and Ling H 1997 Joint time-frequency ISAR using adaptive processing *IEEE Trans. Antennas Propag.* **45** 221

[21]  Borden B 1997 An observation about radar imaging of re-entrant structures with implications for automatic target recognition *Inverse Problems* **13** 1441

[22]  Erdélyi A, Magnus W, Oberhettinger F and Tricomi F G 1954 *Tables of Integral Transforms* vol II (New York: McGraw-Hill)

[23]  Rooney P G 1980 On the $\mathcal{Y}_\nu$ and $\mathcal{H}_\nu$ transformations *Can. J. Math.* **32** 1021

[24]  Stratton J A 1941 *Electromagnetic Theory* (New York: McGraw-Hill)

# 7

---

# Three-Dimensional Imaging

Like HRR object estimation, ISAR imaging cannot yield complete cross-range information about the target (since $\overline{\rho}$ is estimated, and not $\rho$ itself). While ISAR offers a more complete image than that available from HRR profiles (which estimate $\overline{\overline{\rho}}$), it is still insufficient for fully determining the target's orientation in space. Moreover, the cross-range dimension in an ISAR image is usually scaled by Doppler frequency shift and it is often difficult to correctly relate this to the down-range target dimension (scaled in meters). The result is target-image distortion that can adversely affect follow-on image analysis (such as target identification). As we have seen, ISAR imaging can also be computationally intensive and may require long dwell times for data collection†. Unfortunately, alternatives to ISAR imaging which yield clues to the target's cross-range structure—and which also address ISAR's limitations—are few in number. One approach, which appears to hold some promise, relies on a monopulse radar's ability to determine the angular position of a target with great accuracy.

As was discussed in section 2.7, for tracking purposes, interferometric methods are used to determine phase-front normals of the scattered field and this information is used to estimate target bearing. (A 'phase-front' is a surface of constant phase.) The phase-front normals lie parallel to the phase gradient vector and the standard approach in examining angle-of-arrival (AOA) behavior is to study the components of the phase gradient [1, 2]. When a target is composed of point scattering centers, with only one scatterer per range resolution cell (so that, in the time domain, only one scattering center is illuminated at any given instant), then it is sometimes possible to determine the angular position of *each target subelement*. This information can be used to augment the HRR profile, and a three-dimensional image of the target may be generated.

Of course, the ability of a radar to accurately measure the AOA of scattering elements of a finite size target will be range-limited. In addition, when two or

---

† Typical radar systems scan over a range of directions. The *dwell* time is the duration that the radar 'looks' in a specific direction.

more scattering centers are illuminated at the same time, scintillation effects can dominate the AOA measurements and adversely affect the resultant image (this is illustrated in figure 7.1).

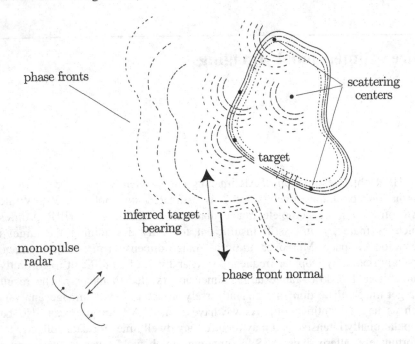

**Figure 7.1**  Scintillation's effect on monopulse tracking. Target subscatterer interference can distort the scattered field phase-front and its normals may point away from the target bearing.

## 7.1  ANGLE TRACKING, SCINTILLATION AND GLINT

Scintillation effects in radar target tracking systems were observed very early in radar development. Known as 'glint', these large and rapid variations in a target's apparent angular position were discovered to be highly correlated with deep-fading amplitude events in the scattered field (i.e. $|H(k, \theta_{\text{fade}})| \approx 0$). But there have also been observations of 'spikes' which did not correspond to deep-amplitude fades, and large glint errors were occasionally seen to be related to small-amplitude fades [3–13].

Glint modeling efforts were begun in an attempt to explain this complex behavior. Typically, such models create an $n$-point representation of a target and use this model to estimate the scattered field. Sometimes, an *n-shape* representation is used. The simplest of these models use a configuration of scattering centers which rotates as a rigid body and whose component number is independent of target aspect. Simple models of this kind are relatively easy to generate and can account for the gross behavior of scattered-field glint— they correctly associate glint spikes with deep-amplitude fades. More complex models try to keep track of the aspect dependence of target scattering centers by assigning multiple $n$-point ($n$-shape) configurations to the same target and applying an 'on/off' decision for which configuration will be used at a given target orientation. Surprisingly, these more complex models often do not fare any better than their simpler progenitors in predicting those glint spikes which are associated with weak fades, or spikes which are not associated with amplitude fades at all. In addition, multi-aspect $n$-point models can be computationally intensive, and $n$-shape models are often prohibitively so.

Within the context of weak far-field targets, however, there is a model which correctly displays the known properties of glint. This model is little-used but is ideally suited to our present needs [14, 15]. For illustration purposes, we will develop the equations for a two-dimensional radar—that is, we will only consider the angle tracking components that lie in the $x$–$y$ plane. It must be understood, however, that equivalent analysis can be performed for angle components in the $y$–$z$ plane.

In the small-angle approximation we can write equation (3.4–3) (also equation (5.2–1)) as the iterated integral

$$H(k_x) = \int_{x_1}^{x_2} \tilde{\rho}_k(x) e^{-ik_x x}\, dx, \quad k_x \in \mathbb{R} \tag{7.1–1}$$

where

$$\tilde{\rho}_k(x) \equiv \int_{\mathbb{R}} \overline{\rho}(x, y') e^{-ik_y y'}\, dy' \tag{7.1–2}$$

and each integral has limits appropriate to the target support (cf figure 7.2). (In particular, $\tilde{\rho}_k(x_1), \tilde{\rho}_k(x_2) \neq 0$.)

If the support interval $(x_1, x_2)$ is finite then $\tilde{\rho}_k$ is a space-limited function and, according to theorem 4.4.1, $H(k_x)$ is an analytic function. Equation (7.1–1) can be used to continue $H(k_x)$ into the complex $w$-plane by defining

$$H(w) \equiv \int_{x_1}^{x_2} \tilde{\rho}_k(x) e^{-iwx}\, dx, \quad w \in \mathbb{C}. \tag{7.1–3}$$

$H(w)$ is an entire function and, consequently, can be represented everywhere by a Taylor series expansion in powers of $w$. It will be more convenient, however, for us to rewrite the Taylor polynomial in terms of the complex zeros $\{w_j\}$ of $H(w)$. Particularly convenient is an expression due to Titchmarsh [16]:

*Theorem* 7.1.1.   Let $H(w)$ be represented by equation (7.1–3) where $x_1$ and $x_2$ define the boundaries of the support of the integrable function $\tilde{\rho}_k(x)$. Then the

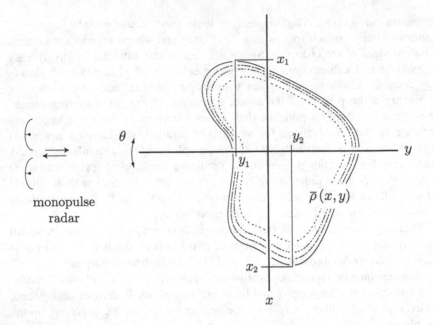

**Figure 7.2**    Target geometry.

product

$$H(w) = H(0) \exp\left(-i\frac{x_1 + x_2}{2}w\right) \prod_{j=1}^{\infty}\left(1 - \frac{w}{w_j}\right) \qquad (7.1\text{-}4)$$

is a conditionally convergent representation of $H(w)$.

On the real-$w$ axis equation (7.1–4) becomes

$$H(k_x) = H(0) \exp\left(-i\frac{x_1 + x_2}{2}k_x\right) \prod_{j=1}^{\infty}\left(1 - \frac{k_x}{w_j}\right). \qquad (7.1\text{-}5)$$

This product represents the scattered field as a function of the *real* variable $k_x = 2k\theta$ in factors of the *complex* zeros of the field continued into the complex plane.

The quadrature phase $\varphi_k(k_x)$ of the scattered field obeys $H(k_x) = |H(k_x)| \exp(i\varphi_k(k_x))$. The derivative of this phase can be found from

$$\frac{\partial \varphi_k(k_x)}{\partial \theta} = \text{Im}\left(\frac{H^*(k_x)}{|H(k_x)|^2}\frac{\partial H(k_x)}{\partial \theta}\right). \qquad (7.1\text{-}6)$$

Applying equation (7.1–6) to equation (7.1–5) yields

$$g \equiv -\frac{1}{2k}\frac{\partial \varphi_k(k_x)}{\partial \theta}$$

$$= \frac{x_1 + x_2}{2} + \frac{\theta}{2}\frac{\partial}{\partial \theta}(x_1 + x_2) - \mathrm{Im}\left(\sum_{j=1}^{\infty}\frac{1}{2k\theta - w_j}\right). \qquad (7.1\text{-}7)$$

This is the glint error in units of the linear cross-range coordinate at the target.

In developing this result, we have assumed that the location of the zeros $\{w_j\}$ does not depend upon target aspect. Since different subcomponents of a complex target will be excited by the incident field at different target orientations, however, there will usually be a dependence of zero location on $\theta$. Consequently, equation (7.1-7) must be viewed as an approximation which will be valid only over a small range of aspect angles for which the locations of the zeros can be taken as fixed.

Equation (7.1-7) is proportional to the pointing distance away from the origin of the coordinate system fixed within the target and displays all of the behavior associated with glint (for example, it shows the strong correlation between glint spikes and amplitude fades). When $(x_1 + x_2)$ is a slowly varying function of target aspect the second term will be vanishingly small. The first term does not vary with small changes in aspect and $\frac{1}{2}(x_1 + x_2)$ is known as the cross-range tracking 'centroid' of the target. As we have discussed, larger aspect variations will often be accompanied by changes in the value of $(x_1 + x_2)$ and so the centroid will vary in position (this is known as 'bright-spot wander'). If the change in centroid position is rapid enough, then the second term in equation (7.1-7) can take on large transient values and we may observe associated spike-like behavior. Note that since this spike-like behavior is associated with a rapid change in $(x_1 + x_2)$, it will generally separate the domain of $\partial \varphi_k(\theta)/\partial \theta$ into subdomains over which equation (7.1-7) is valid to within the small-angle approximation: in each subdomain, $x_1$, $x_2$ and $\{w_j\}$ can be assumed to be approximately constant.

Within each of these subdomains there is also regular spike-like behavior determined by the third term of equation (7.1-7). This behavior, of course, is due to scintillation and equation (7.1-5) can be viewed as a generalization to equation (6.2-1). The $\{w_j\}$ are defined as the points in the complex $w$-plane at which $H(w)$ vanishes. If we write $w_j = \mathrm{Re}(w_j) + i\,\mathrm{Im}(w_j)$ then

$$\mathrm{Im}\left(\sum_{j=1}^{\infty}\frac{1}{2k\theta - w_j}\right) = \sum_{j=1}^{\infty}\frac{\mathrm{Im}\,(w_j)}{(2k\theta - \mathrm{Re}\,(w_j))^2 + \mathrm{Im}\,(w_j)^2} \qquad (7.1\text{-}8)$$

and we can see that a $w_j$ will make a significant contribution to the sum in (7.1-8) only if its imaginary part is sufficiently small.

These results are illustrated in figure 7.3 which displays the glint error measured from an aircraft target rotating on a fixed pedestal (a Firebee target drone [17]). We have chosen a region that displays the effects of all three terms of equation (7.1-7). At aspects less than about 0°, the error is a combination of the constant first term and the spikes of the third term. In section 7.3 we will show that the spacing between these spikes is determined by the cross-range extent $(x_2 - x_1)$. At an aspect of about $1°$, the effective cross-range extent undergoes an abrupt shift and the error is dominated by the second term of equation (7.1–7). For aspects greater than this, the target centroid ('constant' first term) has shifted and the regularity of the glint spikes in this new region has also changed slightly.

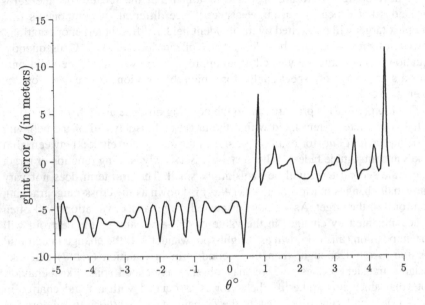

**Figure 7.3**   A plot of glint error as a function of target aspect.

## 7.2   ANGLE-OF-ARRIVAL IMAGING

When $x_1 = x_2$ (for example $\tilde{\rho}_k(x) = A\delta(x)$ in equation (7.1–1)) the third term of equation (7.1–7) will vanish (i.e. there will be no scintillation). If, in addition, $\theta$ is held constant (say, $\theta = 0$), then equation (7.1–7) yields

$$\frac{1}{2k}\frac{\partial \varphi_k(k_x)}{\partial \theta} = x_1. \qquad (7.2–1)$$

This single-scatterer situation is sometimes representative of radar targets illuminated by very narrow time-domain pulses so that each range bin will contain only one scattering center. When such a sparse scattering center situation is true, the target can be imaged in three dimensions by combining range profile information with the $x$ and $z$ scattering center coordinates derived from AOA measurements [18, 19]. Figure 7.4 illustrates this approach.

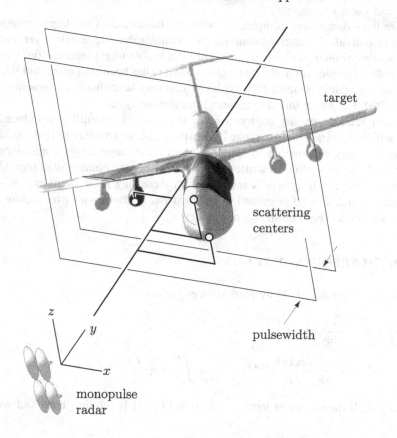

**Figure 7.4**   Three-dimensional angle-of-arrival imaging.

Unfortunately, this method of imaging is highly dependent upon range, and many of the advantages that radar systems hold over optical systems become less compelling when the radar-to-target range is small. High-performance monopulse systems enjoy signal-to-noise ratios of (typically) $10^5$ and this results in an RMS angle measurement error of 0.1 mrad [20]. Consequently, a target with cross-range dimensions of 5 m cannot begin to be imaged using this technique when its range exceeds $R \approx 5/0.0001 \, \text{m} = 50 \, \text{km}$ (*effective* imaging will only occur at much smaller ranges, and $R < 5\text{--}10 \, \text{km}$ seems more reasonable to many). Nevertheless, these techniques may become important in monopulse systems where ISAR imaging is not feasible and high-quality AOA data are already collected for tracking purposes. (An example is the guidance-integrated fuze problem (GIF) in which a radar guided missile uses its existing tracking functionality to locate a target's most vulnerable structures in the (short-range) end-game.)

When the target is very complex, or when the bandwidth is not large enough to resolve individual scattering centers, then scintillation effects take over and the three-dimensional AOA image may not offer usable target information. Use of ISAR-like Doppler variations to isolate scatterers has been suggested, but this introduces an additional level of complexity that may be difficult to implement in a small (for example missile guidance) monopulse system.

Another related imaging approach recognizes that the scintillation-induced glint itself contains target cross-range information and, in a variety of real-world situations, it may be possible to estimate gross cross-range target dimensions from very dense collections of scattering centers. (Since the number of scattering centers on a target will generally increase with increasing radar frequency, this approach—which will be examined in the next two sections—may be required in very small and very high-frequency systems.)

## 7.3  HIGH-FREQUENCY ZEROS

Integrating equation (7.1–3) by parts (twice) yields

$$
H(w) = \frac{1}{-\mathrm{i}w}\tilde{\rho}_k(x)\mathrm{e}^{-\mathrm{i}wx}\bigg|_{x_1}^{x_2}
$$
$$
+ \frac{1}{w^2}\frac{\mathrm{d}\tilde{\rho}_k(x)}{\mathrm{d}x}\mathrm{e}^{-\mathrm{i}wx}\bigg|_{x_1}^{x_2} - \frac{1}{w^2}\int_{x_1}^{x_2}\frac{\mathrm{d}\tilde{\rho}_k(x')}{\mathrm{d}x'}\mathrm{e}^{-\mathrm{i}wx'}\,\mathrm{d}x'. \tag{7.3–1}
$$

For large $|w|$, the dominant term in equation (7.3–1) is the first term and we have

$$
\lim_{|w|\to\infty} wH(w) = 2\exp\left(-\mathrm{i}\frac{x_1 + x_2}{2}w - \tfrac{1}{2}\ln\left(\tilde{\rho}_k(x_1)\tilde{\rho}_k(x_2)\right)\right)
$$
$$
\times \sin\left[\frac{x_2 - x_1}{2}w - \mathrm{i}\tfrac{1}{2}\ln\left(\frac{\tilde{\rho}_k(x_1)}{\tilde{\rho}_k(x_2)}\right)\right]. \tag{7.3–2}
$$

Let $\{w_j\}$ be the zeros of $H(w)$ arranged in order of increasing modulus. For $\tilde{\rho}_k$ defined as in equation (7.1–1), with $x_1 \neq x_2$, the asymptotic (high-frequency) form of the zeros is obtained from equation (7.3–2) as

$$
w_j = \frac{2\pi j}{x_2 - x_1} + \mathrm{i}\frac{1}{x_2 - x_1}\ln\left(\frac{\tilde{\rho}_k(x_1)}{\tilde{\rho}_k(x_2)}\right) + \epsilon_j \tag{7.3–3}
$$

where $\epsilon_j \to 0$ as $j \to \infty$ [21]. Because of equation (7.3–2), $H(w)$ is known as a 'sine-type' function and has zeros that lie (approximately) on a regular lattice with spacing determined by the support of $\tilde{\rho}_k$. If we write $\tilde{\rho}_k(x) = |\tilde{\rho}_k(x)| \exp(i\zeta_k(x))$ then equation (7.3–3) becomes

$$w_j = \frac{2\pi j + \zeta_k(x_2) - \zeta_k(x_1)}{x_2 - x_1} + i\frac{1}{x_2 - x_1} \ln\left|\frac{\tilde{\rho}_k(x_1)}{\tilde{\rho}_k(x_2)}\right| + \epsilon_j. \qquad (7.3\text{–}4)$$

Equations (7.1–8) and (7.3–4) allow us to make a qualitative description of the zero-induced glint. When $2k\theta = \mathrm{Re}\,(w_j)$, $j = 1, \ldots, \infty$, the third term of equation (7.1–7) adds an additional angular glint error of approximately

$$g_{\text{spike}} \approx -2\frac{x_2 - x_1}{\ln|\tilde{\rho}_k(x_2)| - \ln|\tilde{\rho}_k(x_1)|}. \qquad (7.3\text{–}5)$$

(This approximation assumes that the glint 'spike' is large.) Note that the size of the error spike is determined by the behavior of $\tilde{\rho}_k$ near the end point of the interval $(x_1, x_2)$ and that when $|\tilde{\rho}_k(x_1)| \approx |\tilde{\rho}_k(x_2)|$ these spikes can become quite large. In practice, of course, there will be a finite number of such glint spikes within each subdomain of 'fixed' $x_1$ and $x_2$. If $\Delta\eta$ denotes the number of zeros in the interval $\Delta\theta$, then these zeros will have a density given (approximately) by

$$\frac{\Delta\eta}{\Delta\theta} \approx \frac{k(x_2 - x_1)}{\pi}. \qquad (7.3\text{–}6)$$

This density increases with increasing frequency.

These results are important because they demonstrate that important cross-range target information is available to monopulse systems even when there are multiple scattering centers per range bin. Scintillation effects may prevent us from creating the three-dimensional image of section 7.2, but if we can collect a *set* of monopulse range profiles, then we may be able to estimate the target's cross-range extent $x_2 - x_1$.

## 7.4 STATISTICAL METHODS

We can obtain a high-frequency estimate for $\zeta_k$ (in equation (7.3–4)) by using the method of stationary phase. The extremal values of the integrand's phase in equation (7.1–2) occur at $y_1$ and $y_2$ (see figure 7.2) and we obtain

$$\tilde{\rho}_k(x_j) \approx \overline{\rho}(x_j, y_j)e^{-ik_y y_j}, \qquad j = 1, 2. \qquad (7.4\text{–}1)$$

Consequently
$$\Delta\zeta \equiv \zeta_k(x_2) - \zeta_k(x_1) \approx 2k(y_2 - y_1). \qquad (7.4\text{–}2)$$

When the target aspect $\theta$ is not very accurately known it is still possible to discuss $g(\theta)$ as a function of a random variable. Because of equations (7.3–4) and (7.1–8), however, $\theta$ can be interpreted as being shifted by an amount proportional to $\Delta\zeta$ and so the function $g(\theta)$ will depend on the parameter $\Delta\zeta$ as well. $\Delta\zeta$ depends upon (generally unknown) details of the target's structure and is an example of a so-called 'nuisance parameter': we are not really interested in its value, but it determines the model statistics. Fortunately, it is often possible to eliminate nuisance parameters by making appropriate assumptions about their behavior.

### 7.4.1 Glint Statistics

If $g$ is considered to be a possible value of a random variable $G$, then the statistical properties of glint can be developed using results from random variable theory. The main result that we will require concerns the statistical behavior of the amplitude and phase of a complex-valued random variable with normally distributed real and imaginary parts (see, for example, [22]).

*Theorem* 7.4.1. Let $Z = X + iY$ be a complex-valued random variable whose real and imaginary parts are independent random variables that are normally distributed with variance $\sigma^2$. If we write $Z = |Z|\exp(i\Phi)$, then $\Phi$ and $|Z|$ are independent random variables. $\Phi$ will be uniformly distributed and the probability density for $|Z|$ will obey

$$
p_{|Z|}(z) = \begin{cases} \dfrac{z}{\sigma^2}\exp\left(-\dfrac{z^2}{2\sigma^2}\right), & \text{if } z \geqslant 0 \\ 0, & \text{otherwise.} \end{cases}
$$

(The converse is also true.)

Denote the zero-induced component of a monopulse measurement by $g_w$ (i.e. $g_w$ is due to the last term in equation (7.1–7)). The statistical behavior of the random variable $G$ will be a consequence of the statistics of the unknown aspect angles $\{\theta_i\}$ at which the data are collected and which are also considered to be possible values of a random variable $\Theta$. As we have shown, the statistics of $G$ will also depend on any (nuisance) model parameters. These parameters will not depend on $\theta$ and, according to equation (7.3–4), are $l = x_2 - x_1$, $\Delta\zeta = \zeta_k(x_2) - \zeta_k(x_1)$, and $\Delta\rho = \ln|\tilde{\rho}_k(x_2)| - \ln|\tilde{\rho}_k(x_1)|$. In terms of $\theta$ and these parameters, the $n$th term of $g_w$ can be written as

$$
g_w^{(n\text{th term})} = \text{Im}\left(\frac{l}{2kl\theta - 2n\pi - \Delta\zeta - i\Delta\rho}\right) \qquad (7.4\text{–}3)
$$

where $l \in \mathbb{R}^+$ (by definition), $\Delta\zeta \in \mathbb{R}$ and $\Delta\rho \in \mathbb{R}$.

Consider the case where the range of aspects $\Delta\theta$ is small enough that the target cross-range extent is unchanged over the data set. Then, according to equation (7.3–4), the zeros that occur within this range of aspects will differ from each other by some multiple of $2\pi$ and the other parameters will remain fixed. When there are multiple zeros in the aperture $\Delta\theta$, different measured values of

$G$ may be sampled from $\theta$-regions dominated by different zeros. Because of the regular location of these zeros, however, each measurement can, effectively, be considered to be a sample from the region around a single glint spike (i.e. the sample will be collected over one 'spike period')—this observation is reflected in equation (7.3–6).

Assume, without loss of generality, that the sample is collected from the region around the $n$th zero, and let

$$b = 2kl\theta - 2n\pi - \Delta\zeta - i\Delta\rho. \qquad (7.4\text{–}4)$$

be a realization of a random variable $B$. If we set $b = |b| \exp(-i\psi)$, then the zero-induced glint due to one term of equation (7.1–7) is

$$g_w = \frac{l \sin \psi}{|b|}. \qquad (7.4\text{–}5)$$

The statistics of $G$ can therefore be determined by considering $g(l \sin \psi, |b|)$ to be a function of the two random variables $U = L \sin \Psi$ and $V = |B|$, with respective densities $p_{L \sin \Psi}(u)$ and $p_{|B|}(v)$.

The actual probability density $p_B$ of $B$ will generally depend upon the maneuvering behavior of the target and the properties of the target itself. For analytical convenience, it is here assumed that $|B|$ is Rayleigh distributed with standard deviation $\sigma_\rho$ and that $\Psi$ is uniform on $(-\pi, \pi)$ (equivalently, that $B$ is normally distributed about the origin). Similarly, the probability density $p_L$ of $L$ is also assumed to obey a Rayleigh law with standard deviation $\sigma_l$. These assumptions are sometimes motivated by the central limit theorem which, in this case, would be applied to $\tilde{\rho}_k$ considered as a sum of scattering centers [3–6]. The assumed statistics of $B$ follow from equation (7.1–2), while those of $L \geqslant 0$ follow as the modulus of a normally distributed sum. The Rayleigh models for $p_{|B|}$ and $p_L$ are not universally agreed upon [23] but lead to a convenient model for $p_G(g)$ which is known to represent $G$ pretty well.

Theorem 7.4.1 allows us to conclude that the random variable $U = L \sin \Psi$ is normally distributed with variance $\sigma_l^2$. The probability density for $G$ follows as the quotient of a normal and a Rayleigh distributed random variable. The probability mass in the region $U/V < g$ is given by

$$P_G(G < g) = \int_{\frac{u}{v} < g} p_{L \sin \Psi}(u) p_{|B|}(v) \, du \, dv. \qquad (7.4\text{–}6)$$

The probability density is defined as the derivative $dP_G(G < g)/dg$. We obtain

$$p_G(g) = \frac{\Lambda^2/2}{(\Lambda^2 + g^2)^{3/2}} \qquad (7.4\text{–}7)$$

where $\Lambda^2 = \sigma_l^2/\sigma_\rho^2$.

### 7.4.2 Estimation of $\Lambda^2$

According to equation (7.4–7), the statistics of $G$ depend upon by the parameter $\Lambda^2$ which is the second moment $\sigma_l^2$ of $l = x_2 - x_1$ weighted by the 'strength' factor $\sigma_\rho^{-2}$. This is an important piece of cross-range target information that can be extracted from a set of measured monopulse data.

Given such a set of $N$ measurements $\{g_i\}_{i=1,...,N}$, the value of $\Lambda$ that 'best matches' the probability density of equation (7.4–7) will be the one that maximizes the function

$$\mathcal{M}(\Lambda) = \prod_{i=1}^{N} p_g(g_i) = \prod_{i=1}^{N} \frac{\Lambda^2/2}{(\Lambda^2 + g_i^2)^{3/2}}. \tag{7.4–8}$$

$\mathcal{M}(\Lambda)$ is known as the likelihood function and the $\Lambda$ which maximizes $\mathcal{M}$ is known as a maximum likelihood estimate.

Differentiating $\ln \mathcal{M}$ with respect to $\Lambda$ and setting the result to zero yields the relation

$$\sum_{i=1}^{N} \left( \frac{\Lambda^2}{\Lambda^2 + g_i^2} - \frac{2}{3} \right) = 0. \tag{7.4–9}$$

Each of the terms in this equation is a similarly shaped and oriented function of $\Lambda$ having two zeros located symmetrically about the origin. Consequently, the sum of equation (7.4–9) will also have two symmetric zeros and the magnitude of either zero will be the estimate that we seek. The solution to equation (7.4–9) can be obtained numerically (for example by Newton's method).

In monopulse systems which measure both azimuthal and elevation AOA information, there will be two estimates: $g_x$ and $g_z$. If $\phi$ denotes the polar angle in the $x$–$z$ plane, then an estimate $\Lambda(\phi)$ can be obtained as a solution to equation (7.4–9) using $g = g_x \cos \phi + g_z \sin \phi$. Ideally, as $\phi$ varies over $(0, \pi)$, $\Lambda(\phi)$ will trace out an ellipse:

$$\Lambda(\phi) = \frac{d}{1 + e \cos 2(\phi + a)}. \tag{7.4–10}$$

We can use least-squares estimation (for example) to fit a set of the measurements $\{\Lambda^2(\phi_i)\}_{i=1,...,M}$ to the ideal form of equation (7.4–10) [24–26]. In this way, each range profile $\bar{\rho}(y)$ can be augmented by the additional parameters $a(y)$, $d(y)$ and $e(y)$.

Figure 7.5 is an illustration of this statistical moment-based extension to AOA imaging. The computer generated data used in this reconstruction were the calculated azimuth and elevation phase monopulse echo measurements expected to be reflected from the collection of 50 scattering centers displayed in figure 7.5(a). (These subscatterers were assigned fixed random values of amplitude and phase and constrained to lie at random locations on the target support shown in the inset.) The simulated bandwidth of $\Delta\omega = 2\pi \times 600$ MHz centered on $\bar{\omega} = 2\pi \times 30$ GHz was used to determine 100 data values from

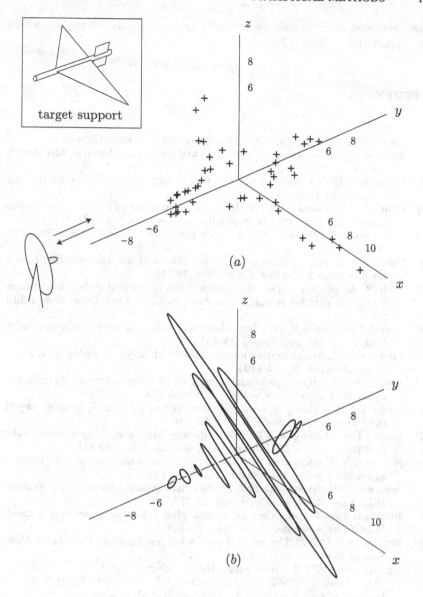

**Figure 7.5** Statistical moment-based image. (*a*) The locations of 50 random amplitude and phase scattering centers fixed on the target support (inset). (*b*) The estimate formed from 100 AOA-enhanced range profiles.

'unknown' target aspects lying within a $2° \times 2°$ aperture. The moment-enhanced range profile calculated from these data is shown in figure 7.5(*b*).

The example of figure 7.5 has been based upon simulated data, although there have been some experimental results (using scale model targets) which also support this analysis [27].

## REFERENCES

[1] Rhodes D R 1959 *Introduction to Monopulse* (New York: McGraw-Hill)
[2] Sherman S M 1984 *Monopulse Principles and Techniques* (Dedham, MA: Artech House)
[3] Delano R H 1953 A theory of target glint or angular scintillation in radar tracking *Proc. IRE* **41** 1778
[4] Dunn J H and Howard D D 1959 The effects of automatic gain control performance on the tracking accuracy of monopulse radar systems *Proc. IRE* **47** 430
[5] Muchmore R B 1960 Aircraft scintillation spectra *IRE Trans. Antennas Propag.* **8** 201
[6] Gubonin N S 1965 Fluctuations of the phase front of the wave reflected from a complex target *Radio Eng. Electr. Phys.* **10** 718
[7] Nullin C M and Aas B 1987 Experimental investigation of correlation between fading and glint for aircraft targets *Proc. Radar-87 Conf.* (New York: IEEE) p 540
[8] Sims R J and Graf E R 1971 The reduction of radar glint by diversity techniques *IEEE Trans. Antennas Propag.* **19** 462
[9] Lind G 1968 Reduction of radar tracking errors with frequency agility *IEEE Trans. Aerospace Electr. Syst.* **4** 410
[10] Lind G 1972 A simple approximate formula for glint improvement with frequency agility *IEEE Trans. Aerospace Electr. Syst.* **8** 854
[11] Asseo S J 1974 Effect of monopulse thresholding on tracking multiple targets *IEEE Trans. Aerospace Electr. Syst.* **10** 504
[12] Loomis J M, III and Graf E R 1974 Frequency-agility processing to reduce radar glint pointing error *IEEE Trans. Aerospace Electr. Syst.* **10** 811
[13] Nichols L A 1975 Reduction of radar glint for complex targets by use of frequency agility *IEEE Trans. Aerospace Electr. Syst.* **11** 647
[14] Borden B 1991 Diversity methods in phase monopulse tracking—a new approach *IEEE Trans. Aerospace Electr. Syst.* **27** 877
[15] Borden B 1994 Requirements for optimal glint reduction by diversity methods *IEEE Trans. Aerospace Electr. Syst.* **30** 1108
[16] Titchmarsh E C 1925 The zeros of certain integral functions *Proc. Lond. Math. Soc.* **25** 283
[17] US Air Force 1974 *Radar Signature Measurements of BQM-34A and BQM-34F Target Drones* ASFWC-TR-74-01, AD785219, The Radar Target Scattering Division, 6585th Test Group, Holloman AFB, NM, January 1974
[18] Howard D D 1975 High range-resolution monopulse tracking radar *IEEE Trans. Aerospace Electr. Syst.* **11** 749
[19] Wehner D R 1987 *High Resolution Radar* (Norwood, MA: Artech House)
[20] Skolnik M L 1980 *Introduction to Radar Systtems* 2nd edn (New York: McGraw-Hill)
[21] Cartwright M L 1930 The zeros of certain integral functions *Q. J. Math., Oxford Ser. (1)* **1** 38
[22] Papoulis A 1965 *Probability, Random Variables, and Stochastic Processes* (New York: McGraw-Hill)

[23]  Leonov A I, Fomichev K I, Barton W F and Barton D 1986 *Monopulse Radar* (Dedham, MA: Artech House)
[24]  Borden B 1986 High-frequency statistical classification of complex targets using severely aspect-limited data *IEEE Trans. Antennas Propag.* **34** 1455
[25]  Borden B 1992 Phase monopulse tracking and its relationship to non-cooperative target recognition *IMA Volumes in Mathematics and its Applications, Radar and Sonar, Part 2* vol 39, ed F A Grünbaum, M Bernfeld and R E Blahut (New York: Springer)
[26]  Borden B 1995 Enhanced range profiles for radar-based target classification using monopulse tracking statistics *IEEE Trans. Antennas Propag.* **43** 759
[27]  Mevel J 1976 Procedure de reconnaissance des formes l'ide d'un radar monostatique *Ann. Telecom.* **131** 111

# 8

## Other Methods

In chapters 1 and 2 we argued that two of the principal goals of a radar system are target location and target classification. The first problem is also the original problem addressed by radar and very effective techniques for determining range, range-rate and bearing have been developed. The second problem has been the subject of chapters 3–7 and our approach has been to examine appropriate (and 'traditional') imaging methods. There are, of course, some not-so-traditional imaging approaches that we have not yet considered. In addition, there are a variety of radar-based techniques that can be applied to the problem of target classification without requiring the intermediate step of image formation. Some of these will be briefly examined here.

### 8.1 RESONANT-FREQUENCY POLES

In chapter 7 we examined the behavior of the scattered field for complex values $w$ of the frequency. This approach turned out to be very useful in describing the properties of glint and scintillation under the weak, high-frequency scatterer model (i.e. $\mathrm{Re}\, w \to \infty$). It turns out that the complex frequency approach is also useful in more general situations—in fact, it can be used to describe general solutions to the magnetic field integral equation (3.1–8) for the currents induced by an incident field on the surface $\partial D$ of a perfect conductor:

$$J(x, t) = \hat{n} \times H_{\text{inc}}(x, t) + \hat{n} \times \frac{1}{4\pi} \int_{\partial D} \mathcal{L}_r \{J\}(x', t') \times \hat{r}\, \mathrm{d}S' \qquad (3.1\text{–}8)$$

where $x, x' \in \partial D$, $\hat{n}$ is the unit normal to $\partial D$ at $x$, $r = |x - x'|$, $\hat{r} = (x - x')/r$ and $\mathcal{L}_r \{J\}(x, t) \equiv \left(r^{-2} + (rc)^{-1}\partial/\partial t\right) J(x, t)$.

Traditional solutions of equation (3.1–8) start out by considering the low- and high-frequency limits. This approach was begun in chapter 3 when we (effectively) looked for solutions to this equation that were of the form $J(x, t) = j(x, \omega)\exp(-i\omega t)$ and examined their behavior in the weak scatterer situation and as $\omega \to \infty$. That analysis led to the basic radar imaging equation (3.4–3) but, in contrast to this imaging approach, several target identification efforts have been devised from *exact* solutions to (3.1–8) and we will now consider these *resonant region* methods.

The singularity expansion method (SEM) was originally developed as a representation of solutions to the scattered field in terms of its singularities in the complex frequency plane [9–12]. This expansion is usually not very useful for direct scattering problems but the representation is very general and *appears* to be ideally suited for application to the problem of target identification. The idea can be stated in a simple way: the complex resonance frequencies of the natural current modes are a consequence of the target's structural geometry and so target identification might be achieved by comparing the resonances of an unknown target with a library of resonances from known targets. The location of these resonances in the complex plane does not depend upon target orientation and this is a feature which makes them very attractive as potential target classifiers.

Consider solutions of the form $J(x, t) = j(x, w) \exp(-iwt)$, where $w \in \mathbb{C}$. Equation (3.1–8) becomes

$$j(x, w) = \tfrac{1}{2} j_{po}(x, w) + \hat{n} \times \frac{1}{4\pi} \int_{\partial D} \left( \frac{1}{r^2} - \frac{iw}{rc} \right) j(x', w) \times \hat{r} \, dS' \quad (8.1\text{–}1)$$

where

$$j_{po}(x, w) = 2\hat{n} \times \int_{\mathbb{R}} H_{inc}(x, t') e^{iwt'} \, dt' \quad (8.1\text{–}2)$$

is the (two-sided Laplace) transform of $J_{po}(x, t)$. Equation (8.1–1) is a second-kind integral equation which we will write in the form

$$(I - F_w) j = \tfrac{1}{2} j_{po} \quad (8.1\text{–}3)$$

for notational convenience.

Let $\lambda_{i;w}$ denote the eigenvalues of $F_w$. According to theorem 4.2.1, $(I - F_w)^{-1}$ exists when $\lambda_{i;w} \neq 1$ for all $i$. As $w$ varies throughout the complex plane, however, there will generally be values $w = w_n$ for which $\lambda_{i;w_n} = 1$ for some $i$. When this occurs, the corresponding term in the eigen expansion for $(I - F_{w_n}) j$ will vanish and the solution of equation (8.1–3) will not be unique (cf equation (A–3)). Evidently, $j$ is a meromorphic function of $w$ with poles at $w_n$. Solutions of the homogeneous equation

$$(I - F_{w_n}) j_n = 0 \quad (8.1\text{–}4)$$

are known as *resonances*.

Since $j$ is a meromorphic function, it can always be represented in a region near one of the poles by a Laurent expansion. A (sometimes) more useful representation, however, is based on the Mittag–Leffler theorem.

*Theorem* 8.1.1 *Mittag–Leffler.*  Suppose that the set of poles $\{w_n\}_{n=1,2,...}$ has no finite accumulation point and let $\{C_{mnp}\}_{p=1,2,...,P_n}$ be a finite number of non-zero constants. Then the principal part of any meromorphic function with poles $\{w_n\}$ has the form

$$j(w) = j_{\text{ent}}(w) + \sum_{m,n=1} \sum_{p=1}^{P_n} \frac{C_{mnp}}{(w - w_n)^p} \, j_{mn} \qquad (8.1\text{--}5)$$

where $j_{\text{ent}}$ is an analytic function in the entire complex plane and the index $m$ accounts for degenerate solutions to equation (8.1–4).

The $C_{mnp}$ are called 'coupling' coefficients while the term $j_{\text{ent}}$ represents a solution to equation (8.1–3) at complex frequencies $w$ for which $(I - F_w)^{-1}$ exists. Observe that since $[r^{-2} - iw(rc)^{-1}]$ is not square-integrable (it has a weak singularity at $r = 0$) we cannot use the usual techniques from standard Fredholm theory (i.e. equation (A–4)) to determine the details of the relationship (8.1–5). Fortunately, our current discussion will not require us to develop an expression for $j_{\text{ent}}$ and so we will only outline an argument, due to Marin [8], based on the observation that the operator $(I - F_w^2) = (I + F_w)(I - F_w)$ *is* square-integrable and which examines the behavior of the equation

$$(I - F_w^2)j = \tfrac{1}{2}(I + F_w)j_{\text{po}} . \qquad (8.1\text{--}6)$$

While the kernel of equation (8.1–1) is not square-integrable, it turns out that $F_w$ is compact when $\partial D$ is smooth (recall equations (3.1–9) and (3.1–10)). From the definition of the inner product (and some vector algebra) we have

$$\langle F_w j, h \rangle$$

$$= \int_{\partial D \times \partial D} \{ L(r, w) \, \hat{n}'' \times [j(x', w) \times \hat{r}] \} \cdot h^*(x'', w) \, \mathrm{d}S' \, \mathrm{d}S'' \qquad (8.1\text{--}7)$$

$$= \int_{\partial D \times \partial D} j(x', w) \cdot \{ L^*(r, w) \, \hat{r} \times [h(x'', w) \times \hat{n}''] \}^* \, \mathrm{d}S'' \, \mathrm{d}S'$$

where $r = |x'' - x'|$, $L(r, w) = (4\pi)^{-1} [r^{-2} - iw(rc)^{-1}]$, $\hat{n}'' = \hat{n}(x'')$ and $\hat{r} = (x'' - x')/r$. The adjoint operator is required to obey $\langle j, F_w^\dagger h \rangle = \langle F_w j, h \rangle$ and, comparing with equation (8.1–7), we can identify

$$F_w^\dagger h \quad \Leftrightarrow \quad - \int_{\partial D} L^*(r, w) \, \hat{r} \times [h(x'', w) \times \hat{n}''] \, \mathrm{d}S'' . \qquad (8.1\text{--}8)$$

If we take $h(x'', w)$ to be the magnetic field, then this last result becomes

$$F_w^\dagger h_{\text{tot}} \quad \Leftrightarrow \quad - \int_{\partial D} L^*(r, w) \, j(x', w) \times \hat{r} \, \mathrm{d}S' \qquad (8.1\text{--}9)$$

where we have used the perfect conductor boundary condition $J(x, t) = \hat{n} \times H_{\text{tot}}(x, t)$ of equation (3.1–1).

From equations (8.1–9) and (3.1–7) we obtain $F_w^{\dagger*} h_{\text{tot}} = -h_{\text{scatt}}$ and consequently

$$(I + F_w^{\dagger*}) h_{\text{tot}} = h_{\text{inc}}. \tag{8.1–10}$$

The *interior* problem is obtained from this by choosing the *inward pointing* surface normal (i.e. $\hat{n} \to -\hat{n}$ in equations (8.1–1)–(8.1–9)). The interior problem is described by

$$(I + F_w) j^{(\text{int})} = -\tfrac{1}{2} j_{\text{po}} \quad \text{and} \quad (I - F_w^T) h_{\text{tot}}^{(\text{int})} = h_{\text{inc}} \tag{8.1–11}$$

where $F_w^T = F_w^{\dagger*}$.

Obviously, any solution to equation (8.1–3) will also solve equation (8.1–6). The converse is not necessarily true, however, but since $F_w$ and $F_w^T$ have the same spectrum then the operators $(I - F_w)$ and $(I - F_w^T)$ will have inverses for the same values of $w$ (and will not have inverses for the same values $w_n$). Moreover, the equations for the interior problem are related through $j^{(\text{int})} = -\hat{n} \times h_{\text{tot}}^{(\text{int})}$ and so $(I - F_w^T)$ and $(I + F_w)$ have inverses for the same values of $w$. Consequently, $(I + F_w)$ and $(I - F_w)$ have inverses for the same values of $w$ and solutions of $(I - F_w^2) j = \tfrac{1}{2}(I + F_w) j_{\text{po}}$ coincide with solutions to equation (8.1–3) for all the values $w$ for which $(I - F_w)^{-1}$ exists. Because of this last observation, we can use equation (A–4) to obtain the solution $j$ for $w \neq w_n$ from equation (8.1–6).

Of course, we are interested in the resonances and it is enough for us that the scattered field of equation (8.1–10) is a meromorphic function with resonant modes determined as source-free solutions to the internal problem. These natural modes are sometimes called 'free-ringing' modes and are distinguished from the 'forced' response modes whose poles are the same as that of the incident excitation and are not target-specific. (The natural modes are akin to the sound a bell makes immediately *after* it has been struck by its clapper.) These resonant solutions are exact—that is, no weak or high-frequency assumptions are required—but, typically, the free-ringing terms will only be excited and persist when the interrogating wave is low frequency and the incident wavelengths are approximately the same as the characteristic target dimensions. The natural resonance frequencies $\{w_n\}$, being source-free solutions to (3.1–8), are aspect independent and this fact is responsible for the belief that the location of the $\{w_n\}$, determined from the scattered field data, might be useful in target recognition strategies. The observation that the resonant region of the scattered field contains terms that depend on source-free solutions to equation (8.1–4) often causes considerable excitement since it holds the promise of aspect-independent target identification—the thinking is that 'source-free' is equivalent to 'aspect-independent'.

To understand why this is not the case, we need to re-examine the important role played by the coupling coefficients in the expansion (8.1–5). The exact nature of these coupling coefficients has been the source of occasional debate and even some confusion. We can obtain a time-domain representation for $J(x, t)$ from equation (8.1–5) by inverse transformation:

$$J(x, t) = \frac{1}{2\pi} \int_C j(x, w') e^{iw't} \, dw' \qquad (8.1–12)$$

where the path of integration is parallel to the real axis and above all of the singularities of $j(x, w)$. Applying this to the representation (8.1–5) results in an expansion whose time-domain coupling coefficients are time-varying. It is known that alternate forms of equation (8.1–5) can be written in which the coupling coefficients are constant in time. Such representations are 'class 1' expansions while (8.1–5) leads to (time-varying) 'class 2' coefficients [14, 15]. The central difference between class 1 and class 2 descriptions is a by-product of the way in which the currents $J_n$ are considered to be 'turned on' by the incident field $H_{inc}$ (which is needed to supply the energy even though the $j_n$ are 'source-free'). After this finite duration driving field has traversed the scatterer body $D$, however, the class 1 and class 2 coefficients become identical.

Regardless of how the coupling coefficients are defined, it follows from (3.1–8) and (8.1–1) that the modes depend upon target aspect, field excitation and field polarization. Consequently, the coupling coefficients *are* aspect dependent since it is certainly not true that all modes will be excited at all aspects. For all practical purposes, SEM must be considered to be an aspect-dependent method.

It is conventional to split the time-domain current into the 'early-time' response and the 'late-time' response. The early-time response is associated with the driven portion of the incident field interaction and the late-time response occurs after the transmitted pulse traverses the target. The late-time portion of the scattered field is necessarily weak in comparison with the driven response. This is particularly true at the high frequencies typically employed by airborne radars, and the effective range at which the free-ringing field can be accurately measured is an issue of current study. Moreover, mid-frequency fields (for which wavelengths are comparable to target size) are often not practicable because conventional systems cannot efficiently generate wavelengths which are much larger than their (usually small) radar antennas [26].

### 8.1.1 Determination of $w_n$ from $H_{scatt}$

It is the late 'free-ringing' response that is most often considered when discussing SEM-based target recognition, and the development of techniques for estimating the poles from noisy scattered field data accounts for much of the recent research. According to our discussion—especially equation (8.1–10)—the transient response $H(t)$ can be represented as

$$H(t) = \int_C h(w) e^{iwt} \, dw \qquad (8.1–13)$$

and will generally show the effects of resonant (source-free) solutions to the associated internal problem described by equation (8.1–11).

For discretely sampled data, we seek an approximation to $H(t)$ of the form

$$H(t) = \sum_{n=1}^{N} h_n e^{iw_n t} = \sum_{n=1}^{N} h_n \mu_n^t \qquad (8.1-14)$$

where $\mu_n = e^{iw_n}$. If $H(t)$ is the late-time response, then $\{h_n\}$ and $\{w_n\}$ are, respectively, the residues and poles of the scattering system. The goal is to determine the $\mu_n$ and the complication is that the expansion is nonlinear in these parameters. This difficulty can be moderated by a technique known as *Prony's method*.

Consider the equations

$$H(t) = \int_{\mathbb{R}} H(t - t')\delta(t')\,dt' \qquad \text{and} \qquad \mu_n^t = \int_{\mathbb{R}} \mu_n^{t-t'}\delta(t')\,dt'. \qquad (8.1-15)$$

These relationships rely on the Dirac distribution $\delta(t)$ and must be modified when $H(t)$ is a sampled function. Prony's method introduces the auxiliary function $\alpha(t_q)$ and requires

$$H(t_p) = \sum_{q=1}^{N} H(t_p - t_q)\alpha(t_q) \qquad (8.1-16)$$

and

$$\mu_n^{t_p} = \sum_{q=1}^{N} \mu_n^{t_p - t_q}\alpha(t_q). \qquad (8.1-17)$$

Assume that the measured data $H_p = H(t_p)$ are uniformly sampled in time $(t_N, t_{N+1}, \ldots, t_{M-1})$. Then equation (8.1–16) is in the form of $M - N$ linear equations and can be written in matrix-vector notation as $\mathbf{H} = \mathbf{F}\alpha$ where $H_p = H_p$, $F_{pq} = H_{p-q}$, and $\alpha_q = \alpha(t_q)$. This set of equations can be solved for the $N$ components of $\alpha$ when $M = 2N$. (When $M > 2N$ the solutions can be estimated by least-squares.)

Once the $\{\alpha_q\}$ are determined from equation (8.1–16) the $\{\mu_n\}$ can be determined as the roots of the algebraic equation (8.1–17) which can be written as

$$(\mu - \mu_1)(\mu - \mu_2) \cdots (\mu - \mu_N) = 0. \qquad (8.1-18)$$

Numerical solution of equation (8.1–18) (for example, by Muller's method) will yield the (complex) resonant frequencies $\{w_n\}$ which are to be used to classify/identify the target. The linear equation (8.1–14) can be written $\mathbf{H} = \mathbf{M}h$ where $M_{np} = \mu_n^{t_p}$, and $h_n = h_n = h(w_n)$, and can be solved for the set $\{h_n\}$—but this step is often omitted. As a result of Prony's method, the nonlinearity of the systems has been concentrated into the single algebraic equation (8.1–17).

Because pole estimation involves the solution of the first-kind equations (8.1–16) and (8.1–17), the problem is ill-posed and the low signal-to-noise environment of radar 'free-ringing' data acquisition often precludes a straightforward approach [18–31]. In fact, the high signal-to-noise requirements of accurate pole extraction methods are believed to preclude SEM-based target recognition strategies from small airborne systems. In addition, it is not yet clear how many terms in the expansion (8.1–14) must be retained in order to affect 'good' target recognition performance. Largely due to these reasons, airborne-system based target recognition methods have come to rely on the driven portion of the scattered field.

## 8.2  POLARIZATION

In section 3.2 we (effectively) approximated the scattered field by retaining only the first term of the Neumann series expansion for the second-kind magnetic field integral equation (3.1–11) so that $J(x, t) = J_{\text{po}}(x, t)$. This resulted in the 'physical optics' approximation to the scattered field which, as we observed in section 3.4, has the same polarization sense as the incident field.

There are important situations where the scattered field will have a different polarization from the incident field but if we treat the vector data *component-wise* then the consequences to our HRR and ISAR image processing algorithms will be manifest only in the strength of the image elements. (Here, the impact of the simple scattering model approximations on image interpretation is far less important than that considered in sections 6.3–6.5.) Information about the target that could be relevant to the problem of classification and identification is lost to the physical optics field, however, and polarization 'enhancements' are sometimes made to image models.

One of the most straightforward of these enhancements is to include the next term in the Neumann expansion of equation (3.1–11) so that

$$J(x, t) \approx J_{\text{po}}(x, t) + \hat{n} \times \frac{1}{2\pi} \oint_{\partial D} \mathcal{L}_r\{J_{\text{po}}\}(x', t') \times \hat{r}\, dS'. \qquad (8.2\text{–}1)$$

Setting $J_{\text{po}} = 2\hat{n} \times H_{\text{inc}}$ and inserting this approximation into equation (3.1–7) yields (in the far-field)

$$
\begin{aligned}
H_{\text{scatt}}(x, t) \approx{} & \frac{1}{2\pi} \int_{\partial D} \mathcal{L}_R\{H_{\text{inc}}\}(x', t')\, \hat{n}' \cdot \hat{R}\, dS' \\
&+ \frac{1}{(2\pi)^2} \int_{\partial D}\oint_{\partial D} \mathcal{L}_R\left\{\mathcal{L}_r\{H_{\text{inc}}\}(x'', t'')\right\}(x', t')\, (\hat{R} \cdot \hat{n}')(\hat{r} \cdot \hat{n}'') \\
&+ \mathcal{L}_R\left\{\mathcal{L}_r\{\hat{r} \cdot H_{\text{inc}}\}(x'', t'')\right\}(x', t')\, \hat{R} \times \left\{\hat{n}' \times n''\right\}\, dS'\, dS''
\end{aligned}
\qquad (8.2\text{–}2)
$$

where $R = x - x'$ and $r = x'' - x'$. (In establishing this last result we have applied the relevant identities for the triple vector product and used the fact that $\hat{R} \cdot H_{\text{inc}} = 0$.)

Substituting $H_{inc}(\boldsymbol{x}, t) = H_0 \exp[i(k\hat{\boldsymbol{R}}\cdot\boldsymbol{x}+kR-\omega t)]$ into equation (8.2–2) and using the fact that $\mathcal{L}_R \to (Rc)^{-1}\partial/\partial t'$ as $R \to \infty$ yields

$$
\begin{aligned}
H_{scatt}(\boldsymbol{x}, t) \approx{}& -\frac{ik}{2\pi R}\int_{\partial D} H_{inc}(\boldsymbol{x}', t')\,\hat{\boldsymbol{R}}\cdot\hat{\boldsymbol{n}}'\,\mathrm{d}S' \\
&-\frac{1}{(2\pi)^2 R}\int_{\partial D}\oint_{\partial D}\left(\frac{ik}{r^2}+\frac{k^2}{r}\right)\Big[H_{inc}(\boldsymbol{x}'', t'')\,(\hat{\boldsymbol{R}}\cdot\hat{\boldsymbol{n}}')(\hat{\boldsymbol{r}}\cdot\hat{\boldsymbol{n}}'') \\
&\qquad\qquad + (\hat{\boldsymbol{r}}\cdot H_{inc}(\boldsymbol{x}'', t''))\,\hat{\boldsymbol{R}}\times(\hat{\boldsymbol{n}}'\times n'')\Big]\,\mathrm{d}S'\,\mathrm{d}S''
\end{aligned}
\tag{8.2–3}
$$

where we have used $t'' = t' + r/c$.

As in section 3.2, we consider the large wave number limit and evaluate the integrals in equation (8.2–3) by the method of stationary phase. In this method, the only significant contributions to the integral will come from a small patches $\sigma_i$ of $\partial D$ around the specular points $\boldsymbol{x}_{sp}$ (i.e. points at which $\hat{\boldsymbol{R}}\cdot\boldsymbol{x}' = 0$). Assume that the surface is smooth and expand $\boldsymbol{x}'$ and $\boldsymbol{x}''$ in a Taylor series about $\boldsymbol{x}_{sp}$. We choose a local principal-axis $u$–$v$ coordinate frame for which $\hat{\boldsymbol{R}}\cdot\hat{\boldsymbol{u}} = 0$ and $\hat{\boldsymbol{R}}\cdot\hat{\boldsymbol{v}} = 0$ (see figure 3.3). Since $r = \boldsymbol{x}'' - \boldsymbol{x}'$, then we can use equation (3.2–4) to write (lowest-order terms only)

$$
\boldsymbol{r}\cdot\hat{\boldsymbol{n}}'' \approx \tfrac{1}{2}\left(\kappa_u u^2 + \kappa_v v^2\right) \quad\text{and}\quad \boldsymbol{r}\cdot\hat{\boldsymbol{H}}_0 \approx -u(H_0\cdot\hat{\boldsymbol{u}}) - v(H_0\cdot\hat{\boldsymbol{v}}) \tag{8.2–4}
$$

where $\kappa_u$ and $\kappa_v$ are the surface principal curvatures at $\boldsymbol{x}_{sp}$. In addition, Rodrigue's formula [30] yields

$$
\hat{\boldsymbol{n}}'' \approx \hat{\boldsymbol{n}}' - \kappa_u u\hat{\boldsymbol{u}} - \kappa_v v\hat{\boldsymbol{v}}. \tag{8.2–5}
$$

The total (round-trip) phase for the second integral in equation (8.2–3) is determined by a phase path consisting (sequentially) of the legs: $\boldsymbol{x} \to \boldsymbol{x}''$, $\boldsymbol{x}'' \to \boldsymbol{x}'$, and $\boldsymbol{x}' \to \boldsymbol{x}$ (see figure 8.1). In terms of an origin fixed to the target at range $R$ from the radar, the total travel-distance phase can be written as $2kR + 2k\hat{\boldsymbol{R}}\cdot\boldsymbol{x}' + k\hat{\boldsymbol{R}}\cdot\boldsymbol{r} + kr$. Applying the approximations (8.2–4) and (8.2–5) yields

$$
\begin{aligned}
H_{inc}&(\boldsymbol{x}'', t'')\,(\hat{\boldsymbol{R}}\cdot\hat{\boldsymbol{n}}')(\hat{\boldsymbol{r}}\cdot\hat{\boldsymbol{n}}'') + (\hat{\boldsymbol{r}}\cdot H_{inc}(\boldsymbol{x}'', t''))\,\hat{\boldsymbol{R}}\times(\hat{\boldsymbol{n}}'\times n'') \\
&\approx -\frac{1}{2r}(\kappa_u u^2 - \kappa_v v^2)\left[(H_0\cdot\hat{\boldsymbol{u}})\hat{\boldsymbol{u}} - (H_0\cdot\hat{\boldsymbol{v}})\hat{\boldsymbol{v}}\right](\hat{\boldsymbol{R}}\cdot\hat{\boldsymbol{n}}') \\
&\quad\times \exp\left(\frac{ik}{2}(\kappa_u u^2 + \kappa_v v^2)\right) e^{ikr}e^{i(2kR+2k\hat{\boldsymbol{R}}\cdot\boldsymbol{x}'-\omega t')} \\
&+ \text{terms odd in } u \text{ and } v.
\end{aligned}
\tag{8.2–6}
$$

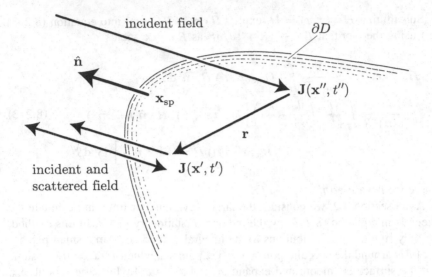

**Figure 8.1**    Multiple scattering geometry.

If $x''$ is restricted to lie within the small patch $(u, v) \in (-a, a) \times (-a, a) \subset \partial D$, then the odd terms will vanish under integration. Inserting the approximation of equation (8.2–6) into equation (8.2–3) and integrating over this patch yields

$$H_{\text{scatt}}(\boldsymbol{x}, t) \approx -\frac{ike^{i(2kR-\omega t)}}{2\pi R} \sum_m A_m e^{i2k\hat{R}\cdot\boldsymbol{x}_m}$$
$$\times \left\{ \boldsymbol{H}_0 - B_m \left[ (\boldsymbol{H}_0 \cdot \hat{u}_m)\hat{u}_m - (\boldsymbol{H}_0 \cdot \hat{v}_m)\hat{v}_m \right] (\kappa_{u;m} - \kappa_{v;m}) \right\}$$

(8.2–7)

where the sum is over all specular points $\boldsymbol{x}_m$, $A_m$ is defined as the stationary phase integral in equation (3.2–5), and $B_m$ is defined as the remaining integral of equation (8.2–3).

The approximation of equation (8.2–7) is sometimes called the 'first-order' correction to physical optics, although it is also based on a high-frequency (stationary phase) approximation to the integrals of equation (8.2–2). This result, which was first obtained by Bennett *et al* [32], is known to accurately represent the depolarization of interrogating signals when the surface is smooth. The depolarizing *behavior* of non-smooth surfaces can be estimated from equation (8.2–7) by considering the limit as $\kappa_{u;m}$ or $\kappa_{v;m}$ becomes infinite. Of course, this limit cannot be justified by the above analysis and, generally, the coefficients $A_m$ and $B_m$ will require a new interpretation—much in the same way as the generalized scatterer density function $\rho_{k,\hat{R}}$ of equation (3.4–1) was a reinterpretation of $A_m$ in the weak scatterer case. Examples of non-smooth depolarizing target structures include edges and corner reflectors. Edges, in particular, are easy to understand since their geometry will not support induced surface currents in directions orthogonal to the edge itself.

The standard approach to characterizing the depolarization effects of a radar target begins by forming the (polarization) scattering matrix $\mathbf{S}$. Let the interrogating wave have polarization $\mathbf{H}_{0;\hat{1}} = H_0\hat{1}$, where $\hat{1}$ denotes a basis polarization vector. Then the 1-1 element of $\mathbf{S}$ is defined by

$$S_{11} = H_0^{-1}\hat{1} \cdot \mathbf{H}_{\text{scatt};\hat{1}}.\qquad(8.2\text{--}8)$$

Similarly, if $\hat{2}$ is a basis polarization vector orthogonal to $\hat{1}$ and $\mathbf{H}_{0;\hat{2}} = H_0\hat{2}$, then

$$S_{22} = H_0^{-1}\hat{2} \cdot \mathbf{H}_{\text{scatt};\hat{2}}.\qquad(8.2\text{--}9)$$

Equations (8.2–8) and (8.2–9) define the so-called 'co-polarization' components of the scattered field. The cross-polarization components are defined by

$$S_{12} = H_0^{-1}\hat{1} \cdot \mathbf{H}_{\text{scatt};\hat{2}}, \quad\text{and}\quad S_{21} = H_0^{-1}\hat{2} \cdot \mathbf{H}_{\text{scatt};\hat{1}}.\qquad(8.2\text{--}10)$$

(Usually, the system is reciprocal and $S_{21} = S_{12}$.)

The basis polarizations $\hat{1}$ and $\hat{2}$ can denote linear, circular or general elliptical polarizations. For purposes of target characterization, the type of polarization is usually chosen to exploit some feature behavior of the target. For example, when $\hat{1}$ and $\hat{2}$ are horizontal and vertical linear polarization, respectively, then the scattering matrix associated with the smooth-surface high-frequency scatterer of equation (8.2–7) is given by

$$S_{\text{HH}} = \frac{1}{2\pi}\sum_m A_m e^{i2k\hat{R}\cdot x_m}\left\{1 - B_m(\kappa_{u;m} - \kappa_{v;m})\cos 2\phi_m\right\}$$

$$S_{\text{VV}} = \frac{1}{2\pi}\sum_m A_m e^{i2k\hat{R}\cdot x_m}\left\{1 + B_m(\kappa_{u;m} - \kappa_{v;m})\cos 2\phi_m\right\}\qquad(8.2\text{--}11)$$

$$S_{\text{HV}} = -\frac{1}{2\pi}\sum_m A_m e^{i2k\hat{R}\cdot x_m}\left\{B_m(\kappa_{u;m} - \kappa_{v;m})\sin 2\phi_m\right\}$$

where $\hat{h} \cdot \hat{u}_m = \cos\phi_m$, $\hat{h} \cdot \hat{v}_m = \sin\phi_m$ and, as has been our convention, we have suppressed the $-R^{-1}ik\exp i(2kR - \omega t)$ prefactor.

The example result of equation (8.2–11) can be interpreted as an extension to the basic scattering equation (3.4–3) upon which the one- and two-dimensional imaging methods of chapters 4 and 5 have been based. In general, all of the results developed for HRR and ISAR imaging can be applied to each of the matrix elements of $\mathbf{S}$. The depolarization properties of the separate image elements can sometimes be determined and this information can, in principle, be applied to the problem of target classification and identification.

Polarization information has traditionally been cataloged with the aid of the Poincaré sphere $P$ whose surface consists of the collection of points which identify all possible types of elliptical polarization. The poles of this sphere represent left- and right-hand circular polarization while the equator represents all possible linear polarization states. General elliptical polarization can be described by two parameters: ellipticity $\epsilon$ (the ratio of the semi-axes of the polarization ellipse) and orientation $\alpha$ (the angle that the semi-major axis make with the horizontal polarization axis). A point on the Poincaré sphere is defined to have 'longitude' $= 2\alpha$ and 'latitude' $= 2\arctan\epsilon$.

The polarization properties of any electromagnetic field $H$ can be represented as a point $h \in P$ and, since this representation is one-to-one [40], the Poincaré sphere is a geometrical encapsulation of all possible polarization related behavior. This geometrical interpretation is conceptually useful in classifying polarization in terms of regions on a sphere and the scattering matrix $\mathbf{S}$ can be viewed as a mapping from $h_{\text{inc}} \in P$ to $h_{\text{scatt}} \in P$.

The phenomenological classification approach of Kennaugh [41] and Huynen [36] can be understood by considering the class of unitary transformations of $\mathbf{S}$:

$$\mathbf{S}' = \mathbf{U}^\dagger \mathbf{S} \mathbf{U}. \tag{8.2--12}$$

It is always possible to find a scattering matrix $\mathbf{S}'$ that will minimize the co- and cross-polarization components of the scattered field. The 'co-pol nulls' are defined by the requirement

$$\hat{1} \cdot (\mathbf{S}'\hat{1}) = 0 \quad \text{and} \quad \hat{2} \cdot (\mathbf{S}'\hat{2}) = 0. \tag{8.2--13}$$

Similarly, the 'cross-pol nulls' are defined by

$$\hat{2} \cdot (\mathbf{S}'\hat{1}) = 0 \quad \text{and} \quad \hat{1} \cdot (\mathbf{S}'\hat{2}) = 0. \tag{8.2--14}$$

Equation (8.2--13) will generally have two different solutions $\mathbf{U}_{\text{co};a}$ and $\mathbf{U}_{\text{co};b}$ which, according to equation (8.2--12), will be required to obey

$$(\mathbf{U}_{\text{co};a}\hat{1}) \cdot (\mathbf{S}\mathbf{U}_{\text{co};a}\hat{1}) = 0 \quad \text{and} \quad (\mathbf{U}_{\text{co};b}\hat{2}) \cdot (\mathbf{S}\mathbf{U}_{\text{co};b}\hat{2}) = 0. \tag{8.2--15}$$

Similarly, the cross-pol nulls are given as solutions to

$$(\mathbf{U}_{\text{cross};c}\hat{2}) \cdot (\mathbf{S}\mathbf{U}_{\text{cross};c}\hat{1}) = 0 \quad \text{and} \quad (\mathbf{U}_{\text{cross};d}\hat{1}) \cdot (\mathbf{S}\mathbf{U}_{\text{cross};d}\hat{2}) = 0. \tag{8.2--16}$$

This last requirement forces the cross-pol nulls to be orthogonal and to occupy antinodal positions on $P$. Moreover, since the trace of a matrix is invariant under unitary transformation, we have

$$\hat{1}' \cdot (\mathbf{S}\hat{1}') + \hat{2}' \cdot (\mathbf{S}\hat{2}') = \text{constant} \quad \text{for all } \mathbf{U} \tag{8.2--17}$$

where $\hat{1}' = \mathbf{U}\hat{1}$ and $\hat{2}' = \mathbf{U}\hat{2}$. Consequently, $\hat{1}_{\text{co}} \equiv \mathbf{U}_{\text{co};a}\hat{1}$ and $\hat{2}_{\text{co}} \equiv \mathbf{U}_{\text{co};b}\hat{2}$ will lie on opposite sides of a circle on $P$. Symmetry considerations require this circle to be centered on either $\hat{\mathbf{1}}_{\text{cross}} \equiv \mathbf{U}_{\text{cross};d}\hat{\mathbf{1}}$ or $\hat{\mathbf{2}}_{\text{cross}} \equiv \mathbf{U}_{\text{cross};c}\hat{\mathbf{2}}$. The four nulls $\hat{\mathbf{1}}_{\text{co}}$, $\hat{\mathbf{2}}_{\text{co}}$, $\hat{\mathbf{1}}_{\text{cross}}$ and $\hat{\mathbf{2}}_{\text{cross}}$ can therefore be seen to define a 'fork' (Huynen's polarization fork) defined by their location on a great circle of $P$. The shape of this fork—as well as its orientation in space—has been applied to the problem of target classification [36].

It is straightforward to show that

$$\hat{\mathbf{1}}_{\text{co}} = a\left[\mathbf{S}_{22}\hat{\mathbf{1}} + \left(-\mathbf{S}_{12} + i\sqrt{\mathbf{S}_{11}\mathbf{S}_{22} - \mathbf{S}_{12}^2}\right)\hat{\mathbf{2}}\right]$$

$$\hat{\mathbf{2}}_{\text{co}} = a\left[\left(-\mathbf{S}_{12} - i\sqrt{\mathbf{S}_{11}\mathbf{S}_{22} - \mathbf{S}_{12}^2}\right)\hat{\mathbf{1}} + \mathbf{S}_{11}\hat{\mathbf{2}}\right]$$

$$\hat{\mathbf{1}}_{\text{cross}} = \frac{b}{2}\left[2\mathbf{S}_{12}\hat{\mathbf{1}} + \left(\mathbf{S}_{22} - \mathbf{S}_{11} + \sqrt{(\mathbf{S}_{11} - \mathbf{S}_{22})^2 + 4\mathbf{S}_{12}^2}\right)\hat{\mathbf{2}}\right]$$

$$\hat{\mathbf{2}}_{\text{cross}} = \frac{b}{2}\left[\left(\mathbf{S}_{11} - \mathbf{S}_{22} - \sqrt{(\mathbf{S}_{11} - \mathbf{S}_{22})^2 + 4\mathbf{S}_{12}^2}\right)\hat{\mathbf{1}} + 2\mathbf{S}_{12}\hat{\mathbf{2}}\right]$$

(8.2–18)

where $a$ and $b$ are coefficients (depending on the $\mathbf{S}_{ij}$) which will scale the length of the vectors so that they lie on $P$. The Huynen fork, as represented in equation (8.2–18), provides a polarimetric representation of the target in terms of combinations of the $\mathbf{S}_{ij}$. These combinations can serve as an alternate basis for parametric target classification schemes and, to some investigators, offer greater interpretability than the individual $\mathbf{S}_{ij}$ themselves. If, for example, an idealized smooth target consisting of *one* specular point is illuminated by a linear polarized interrogating wave, then the scattering matrix will have elements given by one term of the sum in equation (8.2–11). The co- and cross-pol nulls associated with this target are given by

$$\hat{h}_{\text{co}} = (1 + B\Delta\kappa\cos 2\phi)\hat{h} + \left(B\Delta\kappa\sin 2\phi + i\sqrt{1 - B^2(\Delta\kappa)^2}\right)\hat{v}$$

$$\hat{v}_{\text{co}} = \left(B\Delta\kappa\sin 2\phi - i\sqrt{1 - B^2(\Delta\kappa)^2}\right)\hat{h} + (1 - B\Delta\kappa\cos 2\phi)\hat{v}$$

$$\hat{h}_{\text{cross}} = \frac{B}{2}\Delta\kappa\left[-\sin 2\phi\hat{h} + (1 + \cos 2\phi)\hat{v}\right]$$

(8.2–19)

$$\hat{v}_{\text{cross}} = -\frac{B}{2}\Delta\kappa\left[(1 + \cos 2\phi)\hat{h} + \sin 2\phi\hat{v}\right]$$

where $\Delta\kappa = \kappa_u - \kappa_v$ and we have suppressed all of the scaling coefficients. The reader can examine the limiting cases of $\Delta\kappa = 0$ (no depolarization) and $\phi = 0$ (target orientation) to better understand the behavior of the Huynen fork.

Of course, there are many other combinations of the $\mathbf{S}_{ij}$ that could be used for polarimetric based classification. Radar engineers, for example, find it more useful to represent the incident and scattered fields in terms of power and express the polarimetric behavior by the Stokes parameters:

$$s_1 = (\hat{1} \cdot H^*)(\hat{1} \cdot H) + (\hat{2} \cdot H^*)(\hat{2} \cdot H)$$

$$s_2 = (\hat{1} \cdot H^*)(\hat{1} \cdot H) - (\hat{2} \cdot H^*)(\hat{2} \cdot H)$$

$$s_3 = (\hat{1} \cdot H^*)(\hat{2} \cdot H) + (\hat{2} \cdot H^*)(\hat{1} \cdot H)$$

$$s_4 = i[(\hat{2} \cdot H^*)(\hat{1} \cdot H) - (\hat{1} \cdot H^*)(\hat{2} \cdot H)].$$

$$(8.2-20)$$

If $\mathbf{s}_{inc}$ denotes the vector whose components are the Stokes parameters associated with the interrogating radar wave, and $\mathbf{s}_{scatt}$ denotes the similar vector for the scattered wave, then the Müller matrix is defined by the linear transformation $\mathbf{s}_{scatt} = \mathbf{M}\mathbf{s}_{inc}$. The real-valued components of the $4 \times 4$ Müller matrix can be readily expressed in terms of the polarization scattering matrix [33] and these components have formed the basis of phenomenological theories of radar-based target classification (cf [34–39] and references cited therein).

The utility of polarization diversity in image understanding has been demonstrated in many situations and the typical (current) scheme is to compare and contrast images (usually SAR images) created from different transmitted and received polarizations. The main problem with polarimetric data measurements is not so much a question about their utility but, rather, whether the practical advantage can justify the additional cost and complexity of the systems required to collect them. At present, polarization-diverse systems are being investigated as a means to affect foliage penetration and clutter rejection and, should such systems be justified by these arguments, improved target understanding may be an ancillary benefit.

## 8.3   TARGET STRUCTURE-INDUCED MODULATIONS

Target-induced modulations of $H_{scatt}$ are frequently observed: in fact, the earliest aircraft tracking radars of the 1930s detected effects due to propeller modulation. When the incident interrogating field is scattered from a target that is undergoing periodic motion—in particular, a target substructure that is oscillating—then the echo waveform will generally be modulated by this target motion and we can write

$$H_{mod}(t) = a(t)e^{i\phi(t)} H_{scatt}(t) \qquad (8.3-1)$$

where $H_{scatt}$ is the (unmodulated) scattered field obtained from equation (3.4–1) *in the absence of structure oscillation.* (Equation (8.3–1) has made use of a slight notational perversion, but our interest here is in how the amplitude modulation $a(t)$ and the so-called 'angle' modulation $\phi(t)$ affect our previous scattered field results and so there should be no confusion in what follows.)

Since $a(t)$ and $\phi(t)$ are assumed to be induced by the same physical process, they will both be periodic in time with the same period. Denote this period by $T_p = 2\pi/\omega_p$. Then the modulation factors can be expressed in terms of the Fourier series as

$$a(t) = \sum_{n=-\infty}^{\infty} a_n e^{in\omega_p t} \tag{8.3-2}$$

and

$$\phi(t) = \sum_{n=-\infty}^{\infty} \phi_n e^{in\omega_p t} = \sum_{n=0}^{\infty} \beta_n \cos(n\omega_p t - \psi_n) \tag{8.3-3}$$

where the last equation defines $\beta_n$ and $\psi_n$.

The Fourier coefficients $\{a_n\}$ are given by

$$a_n = \frac{1}{T_p} \int_0^{T_p} a(t') e^{-in\omega_p t'} \, dt' \tag{8.3-4}$$

and, consequently, the Fourier transform $A(\omega)$ of $a(t)$ can be written as

$$
\begin{aligned}
A(\omega) &= \int_{\mathbb{R}} a(t') e^{-i\omega t'} \, dt' \\
&= \sum_{n=-\infty}^{\infty} a_n \int_{\mathbb{R}} e^{i(n\omega_p - \omega)t'} \, dt' = \sum_{n=-\infty}^{\infty} a_n \delta(n\omega_p - \omega).
\end{aligned} \tag{8.3-5}
$$

Similarly, if we employ the identity

$$\exp(iz\cos\theta) = \sum_{k=-\infty}^{\infty} i^k J_k(z) e^{ik\theta} \tag{8.3-6}$$

together with equation (8.3–3) so that

$$
\begin{aligned}
w(t) = e^{i\phi(t)} &= \lim_{N\to\infty} \prod_{n=0}^{N} \exp[i\beta_n \cos(n\omega_p t - \psi_n)] \\
&= \lim_{N\to\infty} \sum_{k_1=-\infty}^{\infty} \cdots \sum_{k_N=-\infty}^{\infty} \left\{ \prod_{n=0}^{N} i^{k_n} J_{k_n}(\beta_n) e^{-ik_n\psi_n} \right\} \\
&\quad \times \exp\left( i\omega_p t \sum_{n=0}^{N} nk_n \right)
\end{aligned} \tag{8.3-7}
$$

then it is straightforward to calculate

$$W(\omega) = \lim_{N\to\infty} \sum_{m=-\infty}^{\infty} \gamma_{m;N} \delta(m\omega_p - \omega) \tag{8.3-8}$$

where

$$\gamma_{m;N} = \sum_{k_1,k_2,\ldots,k_N \in S_m} \prod_{n=0}^{N} i^{k_n} J_{k_n}(\beta_n) e^{-ik_n\psi_n} \tag{8.3-9}$$

and $S_m$ is the set of all $N$-tuples of integers obeying $\sum_{n=0}^{N} nk_n = m$.

Equations (8.3–5) and (8.3–9) show that both $A(\omega)$ and $W(\omega)$ are 'line spectra' with spectral components occurring at frequencies that are integer multiples of $\omega_p$. Consequently, according to equation (8.3–1), the spectrum $\mathcal{F}\{H_{\text{mod}}\}(\omega)$ of the scattered field in the presence of these modulations will be the spectrum $\mathcal{F}\{H_{\text{scatt}}\}(\omega)$ convolved with the spectral lines of $A(\omega)$ and $W(\omega)$.

Such characteristic vibration effects have suggested (to some) that a target vibration 'signature' might be formed and used for target identification in much the same way that sonic signatures are used to identify submarines. Since these modulations are imposed on the scattered field, they will not require the incident field to be wide band and a common approach employs a *single* transmitted frequency with narrow band reception. These bandwidth considerations are significantly relaxed from those necessary for the ordinary imaging methods of chapters 4–7 and make modulation-based target identification attractive in some of the more traditional radar environments (which are typically very narrow band). Of course, while the spectral line positions depend upon $\omega_p$, their strength depends on the scattering details in a very complicated way and there have been no real attempts to extract target information from relative modulation amplitude measurements. To date, all such target identification schemes have been based on empirically developed target libraries against which the measured target spectra are to be matched.

There are two common physical mechanisms for echo modulation by modern aircraft: skin vibration and reflection from rotating elements of the propulsion system.

### 8.3.1  The 'Microphone Effect'

In this situation the surface of the aircraft oscillates as a consequence of target motion, engine vibration, etc. When specular reflection is the dominant scattering mechanism the amplitude modulation factor of equation (8.3–1) can often be treated as approximately constant and the angle modulation can be modeled by

$$\phi(t) = 2kD \cos(\omega_p t) \qquad (8.3\text{--}10)$$

where $D$ is the amplitude of surface vibration. Typically, the frequency $\omega_p$ is in the low audible range ($\omega_p/2\pi \sim 0.02$ to 2 kHz) and $D < 10^{-3}$ m.

When the skin vibration is expressible in terms of a single frequency, the angle modulation term of equation (8.3–8) becomes relatively simple and can be written

$$W(\omega) \propto \sum_{m=-\infty}^{\infty} J_m(2kD)\delta(m\omega_p - \omega). \qquad (8.3\text{--}11)$$

In the small-argument approximation $J_0(\beta) \approx 1$, $J_{\pm 1}(\beta) \approx \mp 1/2$ and $J_{|\pm m|>1}(\beta) \approx 0$, this yields

$$W(\omega) \approx \delta(\omega) + kD\delta(\omega - \omega_p) - kD\delta(\omega + \omega_p). \qquad (8.3\text{--}12)$$

From this last result it can be seen that the energy of the first line of the surface vibration induced modulations will be reduced from the unmodulated return by a factor of $\sim (kD)^2$. (At $\omega \sim 2\pi \times 10$ GHz and $D \sim 10^{-3}$ m, for example, $(kD)^2 \sim 0.01$ ($= -20$ dB)—which makes this a very small radar effect. The strength of this line will increase with increasing frequency, and the effect may be important to so-called 'laser radars'.)

### 8.3.2  Jet-Engine Modulation

The modulation induced by jet-engine turbine blades (JEM) is observed at higher frequencies than that of surface vibrations and $\omega_p/2\pi \sim 1$ to 10 kHz are frequently cited values [43]. JEM is not an all-aspect effect and will only make a significant contribution to the scattered field when radar energy couples to the engine duct. Effective coupling between wave and duct implies 'high'-frequency interrogating radar waves, and ray-optic techniques are sometimes invoked to assert that the modulation mechanism is simple 'chopping' by the engine turbine blades. Frequently, JEM is modeled by an amplitude-only term. This model means that a large fraction of the energy that finds its way into the engine duct will be scattered in a modulated form and, owing to the observations of section 6.5, JEM can be a much larger effect than surface vibration modulation. (In practice, surface vibration modulation is difficult to observe at radar frequencies while JEM is difficult to avoid.)

The fundamental observable in JEM-based target classification is the characteristic frequency $\omega_p$ which depends upon the number of blades in the turbine and the speed at which the turbine rotates. Typically, the JEM-line structure will be periodic and this property allows the blade number to be extracted from the data. Blade number is a useful classification parameter and in high signal-to-noise environments it is often possible to determine the blade count for each of the separate stages of multistage turbine engines. In many situations the JEM signature will be sufficiently unique to affect target classification and has been used with moderate success in a limited number of situations.

A variant of JEM—which obviously includes both amplitude and angle modulation effects—is propeller blade modulation. And, because helicopters are dominated at all aspects by these kinds of modulations, this method for target classification may be especially appropriate to non-fixed wing aircraft.

## 8.4  WIDE BAND RADAR

Target rotation rate is a parameter beyond the control of the radar engineer and so there is very little that can be done to improve cross-range resolution in standard ISAR processing methods. Down-range resolution, on the other hand, is controllable and is generally determined by the bandwidth of the interrogating signal. The radar mapping equation (2.5–5) relates the target reflectivity density to the radar measurement through the narrow band ambiguity function and chapter 2 made use of the narrow band approximation to develop many of its results. When very good range resolution is called-for—so that the bandwidth is a large fraction of the center frequency of the carrier wave (for example, $\beta/\omega_0 \sim 0.25$)—then some of these narrow band results must be modified [44–46].

For a signal $s(t)$ with Fourier transform $S(\omega)$ (see equation (2.1–1)), denote by $|s(t)|^2$ the 'instantaneous power' at time $t$ and by $|S(\omega)|^2$ the 'power spectrum' at $\omega$. Our generalization of the results of section 2.6 begins by seeking a function $P(t, \omega)$, called the 'time-frequency distribution', which satisfies the marginal relations

$$\int_{\mathbb{R}} P(t, \omega') \, d\omega' = |s(t)|^2, \qquad \int_{\mathbb{R}} P(t', \omega) \, dt' = |S(\omega)|^2 \qquad (8.4\text{–}1)$$

and such that the total energy of the signal is given by

$$E = \int_{\mathbb{R}^2} P(t', \omega') \, d\omega' \, dt'. \qquad (8.4\text{–}2)$$

For stochastic signals, these results reduce to their probabilistic form. For deterministic signals, this function is really an energy density and the term 'distribution'—although partly a holdover from quantum mechanics—is used because the global average (the 'expectation') of a function $g(t, \omega)$ is defined by

$$\langle g \rangle \equiv \int_{\mathbb{R}^2} g(t', \omega') P(t', \omega') \, d\omega' \, dt'. \qquad (8.4\text{–}3)$$

In addition, the local (or 'conditional') mean at a particular time is defined by

$$\langle g \rangle_t \equiv \frac{\int_{\mathbb{R}} g(t, \omega') P(t, \omega') \, d\omega'}{\int_{\mathbb{R}} P(t, \omega') \, d\omega'} \qquad (8.4\text{–}4)$$

(and similarly for the mean at a particular frequency). Equations (8.4–3) and (8.4–4) are the main motivation for seeking $P(t, \omega)$.

The characteristic function $M(\vartheta, \tau)$ is defined to be the expectation of $e^{i(\vartheta t + \tau \omega)}$:

$$M(\vartheta, \tau) \equiv \int_{\mathbb{R}^2} e^{i(\vartheta t' + \tau \omega')} P(t', \omega') \, d\omega' \, dt'. \qquad (8.4\text{–}5)$$

Characteristic functions are convenient because joint moments can be calculated by differentiation

$$\langle t^n \omega^m \rangle = \frac{1}{i^{n+m}} \frac{\partial^{n+m}}{\partial \vartheta^n \partial \tau^m} M(\vartheta, \tau) \bigg|_{\vartheta, \tau = 0} \qquad (8.4\text{–}6)$$

and so, expanding the exponential in equation (8.4–5) in a Maclaurin series, they may be expressed as

$$M(\vartheta, \tau) = \sum_{n=0}^{\infty} \sum_{m=0}^{\infty} \frac{(i\vartheta)^n (i\tau)^m}{n! m!} \langle t^n \omega^m \rangle. \qquad (8.4\text{--}7)$$

For functions $g(t)$ and $G(\omega)$ we have (from Fourier analysis)

$$\langle g \rangle = \int_{\mathbb{R}} |s(t')|^2 g(t') \, dt' = \int_{\mathbb{R}} S^*(\omega') g\left(i\frac{d}{d\omega'}\right) S(\omega') \, d\omega'$$

and $\qquad (8.4\text{--}8)$

$$\langle G \rangle = \int_{\mathbb{R}} |S(\omega')|^2 G(\omega') \, dt' = \int_{\mathbb{R}} s^*(t') G\left(-i\frac{d}{dt'}\right) s(t') \, dt'.$$

Consequently, we can associate $t$ and $\omega$ with the operators $T$ and $W$ according to the rule

$$T, W \to \begin{cases} t, \quad -i\frac{d}{dt} & \text{in the time domain} \\ i\frac{d}{d\omega}, \quad \omega & \text{in the frequency domain.} \end{cases} \qquad (8.4\text{--}9)$$

A function $g(t, \omega)$ of *both* time and frequency may be formally treated in the same way:

$$\langle g \rangle = \int_{\mathbb{R}} s^*(t') \Gamma(t', W') s(t') \, dt'$$

$$= \int_{\mathbb{R}} S^*(\omega') \Gamma(T', \omega') S(\omega') \, d\omega' \qquad (8.4\text{--}10)$$

where $\Gamma(\cdot, \cdot)$ is the operator 'associated' with $g(t, \omega)$. This association is not well defined. For example, equation (8.4–7) allows us to write

$$M(\vartheta, \tau) = \langle e^{i(\vartheta t + \tau \omega)} \rangle \to \int_{\mathbb{R}} s^*(t') e^{i\vartheta t'} e^{i\tau W'} s(t') \, dt'$$

or $\qquad (8.4\text{--}11)$

$$M(\vartheta, \tau) = \langle e^{i(\vartheta t + \tau \omega)} \rangle \to \int_{\mathbb{R}} S^*(\omega') e^{i\vartheta T'} e^{i\tau \omega'} S(\omega') \, d\omega'.$$

Note, however, that we could just as well have chosen $\exp[i(\vartheta t + \tau \omega)] \to \exp[i(\vartheta T + \tau W)]$. An unambiguous procedure for associating $\Gamma(T, W)$ with $g(t, \omega)$ sets [47]

$$\Gamma(T, W) = \int_{\mathbb{R}^2} \gamma(\vartheta', \tau') \phi(\vartheta', \tau') e^{i(\vartheta' T + \tau' W)} \, d\vartheta' \, d\tau' \qquad (8.4\text{--}12)$$

where

$$\gamma(\vartheta, \tau) = \frac{1}{4\pi^2} \int_{\mathbb{R}^2} g(t', \omega') e^{-i(\vartheta t' + \tau \omega')} \, dt' \, d\omega' \qquad (8.4\text{--}13)$$

and $\phi(\vartheta, \tau)$ is a kernel function which must satisfy certain initial conditions if the correct marginals are to be obtained (see below).

In the time domain the operator $\exp(i\tau W)$ is the translation operator (i.e. $\exp(i\tau W)s(t) = \exp(\tau \frac{d}{dt})s(t) = s(t + \tau)$) and substitution yields the general form of the characteristic function to be

$$M(\vartheta, \tau) = \phi(\vartheta, \tau) \int_{\mathbb{R}} s^*(t')e^{i\vartheta t'} s(t' + \tau) \, dt' . \qquad (8.4\text{--}14)$$

Inverting the definition (8.4–5), with $M(\vartheta, \tau)$ given by equation (8.4–14), yields the general form for distributions:

$$P(t, \omega) = \frac{1}{4\pi^2} \int_{\mathbb{R}^3} e^{-i(\vartheta t + \tau \omega - \vartheta t')} \phi(\vartheta, \tau) s^*(t') s(t' + \tau) \, dt' \, d\vartheta \, d\tau . \qquad (8.4\text{--}15)$$

(This is the so-called *Cohen's class* of time-frequency distributions.)

## 8.4.1    The Kernel Function

The properties of the distribution function are direct consequences of the choice of kernel $\phi(\vartheta, \tau)$. This kernel, in turn, cannot be chosen in an arbitrary way and must obey certain integration and transformation criteria. In particular, the kernel determines whether the density is correctly related to the instantaneous energy spectrum [47].

Integration of $P(t, \omega)$ (equation (8.4–15)) with respect to $\omega$ yields

$$\int_{\mathbb{R}} P(t, \omega') \, d\omega' = \frac{1}{2\pi} \int_{\mathbb{R}^3} \delta(\tau')\phi(\vartheta', \tau')e^{i\vartheta'(t'-t)}$$
$$\times s^*(t')s(t' + \tau') \, d\vartheta' \, dt' \, d\tau' \qquad (8.4\text{--}16)$$
$$= \frac{1}{2\pi} \int_{\mathbb{R}^2} e^{i\vartheta'(t'-t)}\phi(\vartheta', 0)|s(t')|^2 \, d\vartheta' \, dt' .$$

If this is to equal $|s(t)|^2$ then

$$\frac{1}{2\pi} \int_{\mathbb{R}} e^{i\vartheta'(t'-t)}\phi(\vartheta', 0) \, d\vartheta' = \delta(t' - t) \quad \Rightarrow \quad \phi(\vartheta, 0) = 1. \qquad (8.4\text{--}17)$$

Similarly, it is easy to show that $\phi(0, \tau) = 1$.

As in chapter 2, we again assume that $s(t)$ is normalized to unit energy. It follows that

$$\int_{\mathbb{R}^2} P(t', \omega') \, dt' \, d\omega' = 1 \quad \Rightarrow \quad \phi(0, 0) = 1. \qquad (8.4\text{--}18)$$

In addition, if we require $P(t, \omega)$ to be real-valued (as is appropriate for a description of energy density), then

$$\phi(\vartheta, \tau) = \phi^*(-\vartheta, -\tau) . \qquad (8.4\text{--}19)$$

Because of equation (8.4–15), the study of Cohen-class density functions is reduced to the study of $\phi(\vartheta, \tau)$—chosen to obey the relations (8.4–17)–(8.4–19). Two common choices are [47]:

Wigner–Ville: $\phi(\vartheta, \tau) = 1 \quad \Rightarrow$

$$P_{WV}(t, \omega) = \frac{1}{2\pi} \int_{\mathbb{R}} e^{-i\omega\tau'} s^*(t) s(t + \tau') \, d\tau'$$

and                                                                                   (8.4–20)

Choi–Williams: $\phi(\vartheta, \tau) = e^{-\vartheta^2\tau^2/\sigma} \quad \Rightarrow$

$$P_{CW}(t, \omega) = \frac{1}{4\pi^{3/2}} \int_{\mathbb{R}^2} \sqrt{\frac{\sigma}{\tau'^2}} e^{-\sigma(t'-t)^2/4\tau'^2 - i\omega\tau'}$$

$$\times s^*(t') s(t' + \tau') \, dt' \, d\tau'.$$

These example distributions are known to have problems that result in artifacts and interpretation difficulties—in particular, they are not strictly positive-valued. The spectrogram

$$P_{spect}(t, \omega) = \left| \frac{1}{2\pi} \int_{\mathbb{R}} e^{-i\omega t'} s(t') h(t' - t) \, dt' \right|^2$$                          (8.4–21)

is obviously strictly positive and is obtained from

$$\phi(\vartheta, \tau) = \int_{\mathbb{R}} e^{-i\vartheta t'} h(t' + \tau) h^*(t') \, dt'$$                          (8.4–22)

where $h(t)$ is a 'window' function. Though widely used, the spectrogram does not yield the correct marginals since satisfying $\phi(\vartheta, 0) = \phi(0, \tau) = 1$ requires that $|h(t)|^2 = \delta(t)$ *and* $|H(\omega)|^2 = \delta(\omega)$, and this is impossible. In practice, $h(t)$ is chosen by either time or frequency resolution requirements.

### 8.4.2   The Wide Band Ambiguity Function

Using equations (2.5–1) and (2.5–3), we can express the signal $s_{scatt}(t)$ due to a waveform reflected from a simple point scatterer as a time delayed transmission signal $s(t)$:

$$s_{scatt}(t) = s(t - t_d) = \sqrt{\alpha}\, s \left( \alpha t - \frac{2R}{c + v} \right)$$                          (8.4–23)

where the target has radial velocity $|v| \ll c$, $\alpha = 1 - 2v/(c + v)$ and we have used $k = \omega/(c + v)$ in the (Galilean) transformed frame. (The scale factor is required for energy conservation.) This result can be written in the form

$$s_{scatt}(t) = \sqrt{D}\, s(D(t + T))$$                          (8.4–24)

where $D = (c - v)/(c + v)$ is known as the 'Doppler stretch factor' and $T = -2R/(c - v)$ is the (scaled) signal delay at $t = 0$.

The cross-correlation of a test pulse $u_m(t) = \sqrt{a}\, s_m(a(t' + t))$ with a point scatterer echo $s_{\text{scatt};n}(t) = \sqrt{D}\, s_n(D(t + T))$ is given by [47–49]

$$\langle s_{\text{scatt};n}, u_m \rangle (t, a) = \int_{\mathbb{R}} \sqrt{D}\, s_n(D(t' + T)) \sqrt{a}\, s_m^*(a(t' + t))\, dt'$$

$$= \sqrt{\frac{D}{a}} \int_{\mathbb{R}} s_n \left( \frac{D}{a}[t' + a(T - t)] \right) s_m^*(t')\, dt'. \tag{8.4–25}$$

If we define the *wide band ambiguity function* by

$$A_{nm}(\tau, \alpha) \equiv \sqrt{\alpha} \int_{\mathbb{R}} s_n(\alpha(t' + \tau)) s_m^*(t')\, dt' \tag{8.4–26}$$

then we can write

$$\langle s_{\text{scatt};n}, u_m \rangle (t, a) = A_{nm} \left( a(T - t), \frac{D}{a} \right). \tag{8.4–27}$$

Note that by expanding $s_n(\alpha(t' + t))$ in powers of $\epsilon = 2v/(c + v)$, and assuming $s(t)$ to be slowly varying with respect to its center frequency, we can express the narrow band ambiguity function (equation (2.6–1)) in terms of $A_{mn}$ as [44–46, 49]

$$\chi_{mn}(t, \vartheta) = \frac{1}{\sqrt{\alpha}} A_{mn}(t/\alpha, \alpha) + o(\epsilon). \tag{8.4–28}$$

The wide band generalization of the 'radar mapping equation' (5.4–2) is based on a target reflectivity density $\varrho(T, D)$ defined in such a way that

$$s_{\text{scatt};n}(t) = \int_{\mathbb{R}^+} \int_{\mathbb{R}} \varrho(T', D') \sqrt{D'} s_n(D'(t + T')) \frac{dT'\, dD'}{D'}. \tag{8.4–29}$$

(Note that $\varrho(T, D)$ is not generally the same function as the $\rho(x)$ developed in chapter 3. $\varrho$ and $\rho$ are, however, closely related: both are appropriate to 'weak scattering' models and $\varrho$ can be considered to result from a wide band extension of $\rho$.) It is easy to show that the reflectivity image can be expressed as

$$\langle s_{\text{scatt};n}, u_m \rangle (t, a) = \int_{\mathbb{R}^+} \int_{\mathbb{R}} \varrho(T', D')$$

$$\times A_{nm} \left( a(T' - t), \frac{D'}{a} \right) \frac{dT'\, dD'}{D'}. \tag{8.4–30}$$

The reflectivity image is generally complex-valued. For many purposes, we are more interested in the intensity image

$$\langle s_{\text{scatt};n}, u_m \rangle^2 (t, a) = \int_{\mathbb{R}^+} \int_{\mathbb{R}} \int_{\mathbb{R}^+} \int_{\mathbb{R}} \varrho(T', D') \varrho(T'', D'')$$

$$\times A_{nm} \left( a(T' - t), D'/a \right) A_{nm}^* \left( a(T'' - t), D''/a \right) \tag{8.4–31}$$

$$\times \frac{dT'\, dD'}{D'} \frac{dT''\, dD''}{D''}.$$

The incoherent point model assumes that the cross-terms in equation (8.4–31) cancel out under integration. This common model implies that

$$\varrho(T, D)\varrho(T', D') = \varsigma(T, D)\delta(T - T', D - D')  \qquad (8.4\text{--}32)$$

where $\varsigma(T, D)$ is the target cross section density. Under this model, equation (8.4–31) simplifies to

$$\langle s_{\text{scatt};n}, u_m\rangle^2(t, a) = \int_{\mathbb{R}^+}\int_{\mathbb{R}} \varsigma(T', D')$$
$$\times \left|A_{nm}\left(a(T' - t), D'/a\right)\right|^2 \frac{\mathrm{d}T'\,\mathrm{d}D'}{D'^2}. \qquad (8.4\text{--}33)$$

(Recall that some of the reasons for the failure of this model—as well as some of their consequences—were the discussed in chapter 6.)

The imaging kernel in equation (8.4–33) is

$$|A_{nm}(t, a)|^2 = a\int_{\mathbb{R}^2} s_n(a(t' + t))s_n^*(a(t' + t + \tau'))$$
$$\times s_m^*(t')s_m(t' + \tau')\,\mathrm{d}t'\,\mathrm{d}\tau'. \qquad (8.4\text{--}34)$$

Let $\omega_0$ denote the 'carrier frequency' of the test function. Comparing equation (8.4–34) with equation (8.4–15) in the form

$$P\left(at, \frac{\omega_0}{a}\right) = \frac{1}{4\pi^2}\int_{\mathbb{R}^3} e^{i\vartheta'(t'-at)}e^{-i\omega_0\tau'/a}\phi(\vartheta', \tau')$$
$$\times s^*(t')s(t' + \tau')\,\mathrm{d}t'\,\mathrm{d}\vartheta'\,\mathrm{d}\tau' \qquad (8.4\text{--}35)$$

allows us to conclude that the kernel associated with the intensity image (under the non-interacting point scatterer model) obeys

$$\int_{\mathbb{R}} \phi(\vartheta', \tau)e^{i\vartheta't}\,\mathrm{d}\vartheta' = \frac{2\pi}{a}s_m\left(\frac{t}{a}\right)s_m^*\left(\frac{t+\tau}{a}\right)e^{i\omega_0\tau/a}. \qquad (8.4\text{--}36)$$

Consequently

$$\phi\left(\frac{\vartheta}{a}, a\tau\right) = \int_{\mathbb{R}} s_m(t')s_m^*(t' + \tau)e^{-i\vartheta t'}e^{i\omega_0\tau}\,\mathrm{d}t' \qquad (8.4\text{--}37)$$

which is the kernel for the spectrogram with the window function set to the envelope of the test function—but with a time scale factor [47].

As we have pointed out, one of the problems with the spectrogram is its dependence on a fixed and predetermined window function for estimating the properties of the signal. If the signal varies slowly in comparison with the extent of the window then the estimate will generally be 'good'. If, however, the signal varies rapidly within the window, then the estimate may prove unreliable. Equation (8.4–36) shows that the so-called *scaleogram* of equation (8.4–34) offers an interesting 'solution' to the problem of fixed-window/resolution limitation. Here, a fixed *shape* window will be time scaled by the factor $a^{-1}$ in such a way that when the signal varies rapidly (i.e., its scale decreases) the window extent will be correspondingly decreased. This idea is at the heart of scale-based signal processing [50, 51].

### 8.4.3   Choice of Signals

Recall from section 2.2 that the Hilbert transform (equation (2.2–1)) can be used to form the imaginary part of a complex signal when only its real part is specified. This choice for the imaginary part yields a complex signal whose spectrum is single-sided. The Hilbert transform is generally not local in time and is impractical to implement in radar systems, but we also observed that when the real part of the radar signal was chosen to be of the form $p(t) = a(t) \cos \Phi(t)$ then the Hilbert transform could be approximated by the quadrature model—provided the spectrum has positive support. Consequently, these practical considerations have led to radar signals which are selected from the Hardy space of functions $H^2(\mathbb{R}) = \{s \in L^2(\mathbb{R}) \mid \mathrm{supp}(\mathcal{F}\{s\}) \subset [0, \infty)\}$. $H^2(\mathbb{R})$ is a closed subspace of $L^2(\mathbb{R})$ [52].

Because our signals $s_n(t)$ are restricted to $H^2(\mathbb{R})$, equation (8.4–29) shows that the measurement space $\mathcal{M}$ must be a subset of $H^2(\mathbb{R})$ (cf the Appendix). Let $\varrho_{\mathcal{M}} \in \mathcal{M} \subset H^2(\mathbb{R})$ denote the projection of $\varrho$ onto the measurement space. The goal of 'ordinary' radar imaging is to recover $\varrho_{\mathcal{M}}(T, D)$ from the measured data, and we need to 'invert' the reflectivity image of equation (8.4–30) in much the same way as we did in the reconstructions of section 4.3, equation (5.2–2) and theorem A.9. That is, we want to be able to express the estimate $\hat{\varrho}_{\mathcal{M}}$ as

$$\hat{\varrho}_{\mathcal{M}}(T, D) = \int_{\mathbb{R}^+} \int_{\mathbb{R}} \langle s_{\mathrm{scatt};n}, u_m \rangle(t', a') \\ \times B_{jk}\left(a'(T - t'), \frac{D}{a'}\right) \frac{dt' \, da'}{a'} \tag{8.4–38}$$

for some $B_{jk}(\tau, \alpha)$.

If we write $B_{jk}(\tau, \alpha)$ in the form

$$B_{jk}(\tau, \alpha) = \sqrt{\alpha} \int_{\mathbb{R}} s_j(\alpha(t' + \tau)) s_k^*(t') \, dt' \tag{8.4–39}$$

then it is straightforward to show that equation (8.4–38) implies that the estimated target reflectivity density is related to the actual density by

$$\hat{\varrho}_\mathcal{M}(T, D) = \int_{\mathbb{R}^+} \frac{S_k(\omega')S_m^*(\omega')}{\omega'} d\omega'$$

$$\times \int_{\mathbb{R}^+} \int_{\mathbb{R}} \varrho_\mathcal{M}(T', D')A_{nj}\left(D(T' - T), \frac{D'}{D}\right) \frac{dT' dD'}{D'}. \qquad (8.4\text{--}40)$$

When $s_k = s_m$ this is a well defined operation provided

$$\int_{\mathbb{R}^+} \frac{|S(\omega')|^2}{\omega'} d\omega' < \infty. \qquad (8.4\text{--}41)$$

(Equation (8.4–41) is known as the *admissibility condition*.)

Equation (8.4–40) also shows that the ideal imaging kernel will be sharply peaked at $(\tau, a) = (0, 1)$ and vanish everywhere else. (Compare this result with the sinc function behavior of equations (4.3–3) and (5.2–2).) There are limits to how well this may be realized, however, and it is well known that matched Gaussians are optimal (in the sense that they minimize the uncertainty relation) for narrow band ambiguity functions. Unfortunately, Gaussians do not belong to $H^2(\mathbb{R})$. More generally, it is known that any good choice for $s(t)$ must obey [52]:

(i) $s \in H^2(\mathbb{R})$;
(ii) $s(t)$ satisfies the admissibility condition (8.4–41).

This allows the quadrature model to be applied to the definition of phase and allows $\varrho_\mathcal{M}(T, D)$ to be decomposed in the $A_{jk}$. (Although image reconstruction based on equation (8.4–38) may not be the most effective approach—see [44].)

## 8.5 FUTURE EFFORTS

A current 'theme' in radar-based target identification (based on intermediate images) is the classification of *target subscatterers*. This categorization is generally done for two reasons: better target models usually result in better images; and more complete descriptions of target elements provide additional cues for target/template matching algorithms. A formal approach towards this kind of taxonomy can be modeled after the analysis of sections 6.4 and 6.5 since, in addition to re-entrant structures, there are various other target subcomponents that are non-pointlike and may lead to radar image artifacts. A partial list of these structures includes: corner reflectors, multiple subresolution point scatterers and surface edges.

The scheme that we used to examine the cavity/inlet behavior (equation (6.5–3)) has an easy generalization:

$$\hat{\rho}(x, y) \approx \int_{\mathbb{R}^2} dx' \, dy' \, \overline{\rho}\left(x', y'\right) \qquad (8.5\text{--}1)$$

$$\times \begin{cases} A_{\text{inlet}}(\boldsymbol{x}, \boldsymbol{x}') & \text{if } \boldsymbol{x}' = \boldsymbol{x}_{\text{inlet}} \\[4pt] A_{\text{corner reflector}}(\boldsymbol{x}, \boldsymbol{x}') & \text{if } \boldsymbol{x}' = \boldsymbol{x}_{\text{corner reflector}} \\[4pt] A_{\text{multiple scatterers}}(\boldsymbol{x}, \boldsymbol{x}') & \text{if } \boldsymbol{x}' = \boldsymbol{x}_{\text{multiple scatterers}} \\[4pt] \quad\vdots \\[4pt] \text{sinc}(\overline{k}\Delta\theta(x - x')) \\[4pt] \times \, \text{sinc}(\Delta k(y - y'))e^{i2\overline{k}(y-y')} & \text{otherwise.} \end{cases}$$

While the consequences of this description have yet to be fully explored, they may be employable in the same fashion as the analysis of section 6.5 to help codify an approach that has (historically) been largely *ad hoc* in the ISAR/HRR literature. At its foundation, this approach is really just a means by which 'model-incorrect' elements can be associated with a generalized, spatially-varying, point-spread function. Once this point-spread function is determined, of course, then its effects on the image can—in principle—be mitigated.

The immediate future will almost certainly see target identification schemes which 'fuze' various imaging and non-imaging techniques. Combining range profiles with JEM, for example, is an obvious marriage since they use non-redundant data sets obtainable by many existing radar systems. Polarization enhancements to any of the imaging methods may also prove useful, although there will usually be some level of data redundancy and the realizable advantages must be examined on a case-by-case basis.

It is clear that the continuing improvement of radar systems (and the computational engines that process their data) will affect the resolution and sensitivity of radar-based target images. But many of these improvements will probably only be evolutionary advances in the sense that they may still be based upon an incorrect scattering model approximation that 'misinterprets' the contributions of non-weak and non-pointlike scatterers. The 'revolutionary' gains that are possible from radar systems with greater bandwidth, greater sensitivity, tailored waveforms and more powerful computational engines, will require more thorough target models and greater care in their application.

# REFERENCES

[1]   Boerner W-M 1985 *Inverse Methods in Electromagnetic Imaging (NATO ASI Series C: Mathematical and Physical Sciences)* vol 143, ed W-M Boerner *et al* (Dordrecht: Reidel)

[2]   Baum C E 1997 Discrimination of buried targets via the singularity expansion *Inverse Problems* **13** 557

[3]   Baum C E 1971 On the singularity expansion method for the solution of electromagnetic interaction problems *Interaction Note 88* (Kirtland AFB, NM: Air Force Weapons Laboratory)

[4]    Baum C E 1976 The singularity expansion method *Transient Electromagnetic Fields* ed L B Felsen (New York: Springer)

[5]    Michalski K A 1981 Bibliography of the singularity expansion method and related chapters *Electromagnetics* **1** 493

[6]    Pearson L W 1983 Present thinking on the use of the singularity expansion in electromagnetic scattering computation *Wave Motion* **5** 355

[7]    Marin L and Latham R W 1972 Representation of transient scattered fields in terms of free oscillations of bodies *Proc. IEEE* **60** 640

[8]    Marin L 1973 Natural-mode representation of transient scattered fields *IEEE Trans. Antennas Propag.* **21** 809

[9]    Colton D and Kress R 1983 *Integral Equation Methods in Scattering Theory* (New York: Wiley)

[10]   Michalski K A 1982 On the class 1 coupling coefficient performance in the SEM expansions for current density on a scattering object *Electromagnetics* **2** 201

[11]   Morgan M A 1984 Singularity expansion representations of fields and currents in transient scattering *IEEE Trans. Antennas Propag.* **32** 466

[12]   Kennaugh E M and Moffatt D L 1965 Transient and impulse response approximations *Proc. IEEE* **53** 893

[13]   Moffatt D L and Mains R K 1975 Detection and discrimination of radar targets *IEEE Trans. Antennas Propag.* **23** 358

[14]   Berni A J 1975 Target identification by natural resonance estimation *IEEE Trans. Aerospace Electr. Syst.* **11** 147

[15]   Baum C E, Rothwell E J, Chen K-M and Nyquist D P 1991 The singularity expansion method and its application to target identification *Proc. IEEE* **79** 1481

[16]   VanBlaricum M L and Mittra R 1975 A technique for extracting the poles and residues of a system directly from its transient response *IEEE Trans. Antennas Propag.* **23** 777

[17]   Chuang C W and Moffatt D L 1976 Natural resonance of radar target via Prony's method and target discrimination *IEEE Trans. Aerospace Electr. Syst.* **12** 583

[18]   VanBlaricum M L and Mittra R 1978 Problems and solutions associated with Prony's method for processing transient data *IEEE Trans. Antennas Propag.* **26** 174

[19]   Brittingham J N, Miller E K and Willows J L 1980 Pole extraction from real frequency information *Proc. IEEE* **68** 263

[20]   Ksienski D A 1985 Pole and residue extraction from measured data in the frequency domain using multiple data sets *Radio Sci.* **20** 13

[21]   Park S-W and Cordaro J T 1988 Improved estimation of SEM parameters from multiple observations *IEEE Trans. Aerospace Electr. Syst.* **30** 145

[22]   Baynard J-P R and Schaubert D H 1990 Target identification using optimization techniques *IEEE Trans. Antennas Propag.* **38** 450

[23]   Kennaugh E M 1981 The K-pulse concept *IEEE Trans. Antennas Propag.* **29** 327

[24]   Chen K-M, Nyquist D P, Rothwell E J, Webb L L and Drachman B 1986 Radar target discrimination by convolution of radar returns with extinction pulses and single-mode extraction signals *IEEE Trans. Antennas Propag.* **34** 896

[25]   Morgan M A 1988 Scatterer discrimination based upon natural resonance annihilation *J. Electromag. Waves Appl.* **2** 481

[26]   Kraus J D 1988 *Antennas* (New York: McGraw-Hill)

[27]   Bojarski N N 1967 Three dimensional electromagnetic short pulse inverse scattering *Special Projects Lab. Report* (Syracuse, NY: Syracuse University Research Corporation)

[28]   Bojarski N N 1982 A survey of the physical optics inverse scattering identity *IEEE*

*Trans. Antennas Propag.* **30** 980

[29] Bleistein N 1989 Large wave number aperture-limited Fourier inversion and inverse scattering *Wave Motion* **11** 113

[30] Stoker J J 1969 *Differential Geometry* (New York: Wiley-Interscience)

[31] Boerner W-M 1980 Polarization utilization in electromagnetic inverse scattering *Inverse Scattering Problems in Optics* ed H P Baltes (New York: Springer) ch 7

[32] Bennett C L, Auckenthaler A M, Smith R S and DeLorenzo J D 1973 Space time integral equation approach to the large body scattering problem *RADC-CR-73-70 AD763794* (Sudbury, MA: Sperry Rand Research Center)

[33] Ishimaru A 1978 *Wave Propagation and Scattering in Random Media* (New York: Academic)

[34] Copeland J R 1960 Radar target classification by polarization properties *Proc. IRE* **48** 1290

[35] Knott E F and Senior T B A 1972 Cross polarization diagnostics *IEEE Trans. Antennas Propag.* **20** 223

[36] Huynen J R 1978 Phenomenological theory of radar targets *Electromagnetic Scattering* ed P L E Uslenghi (New York: Academic)

[37] Manson A C and Boerner W-M 1985 Interpretation of high-resolution polarimetric radar target down-range signatures using Kennaugh's and Huynen's target characteristic operator theories *Inverse Methods in Electromagnetic Imaging* ed W-M Boerner (Dordrecht: Reidel)

[38] Chamberlain N F, Walton E K and Garber F D 1991 Radar target identification of aircraft using polarization-diverse features *IEEE Trans. Aerospace Electr. Syst.* **27** 58

[39] Cloude S R and Pottier E 1996 A review of target decomposition theorems in radar polarimetry *IEEE Trans. Geosci. Remote Sensing* **34** 498

[40] Deschamps G A 1951 Geometrical representation of the polarization of a plane electromagnetic wave *Proc. IRE* **39** 540

[41] Kennaugh E M 1952 Polarization properties of target reflection *Technical Report 389–2* (Griffis AFB, NY)

[42] Collot G 1991 Fixed/rotary wings classification recognition *Proc. CIE 1991 Int. Conf. on Radar (CICR-91) (Beijing, China)* (New York: IEEE) p 610

[43] Bell M R and Grubbs R A 1993 JEM modeling and measurement for radar target identification *IEEE Trans. Aerospace Electr. Syst.* **29** 73

[44] Naparst H 1991 Dense target signal processing *IEEE Trans. Inform. Theory* **37** 317

[45] Kalnins E G and Miller W Jr. 1992 A note on group contractions and radar ambiguity functions *IMA Volumes in Mathematics and its Applications, Radar and Sonar, Part 2* vol 39, ed F A Grünbaum, M Bernfeld and R E Blahut (New York: Springer) p 71

[46] Maass P 1992 Wideband approximation and wavelet transform *IMA Volumes in Mathematics and its Applications, Radar and Sonar, Part 2* vol 39, ed F A Grünbaum, M Bernfeld and R E Blahut (New York: Springer) p 83

[47] Cohen L 1989 Time-frequency distributions - a review *Proc. IEEE* **77** 941

[48] Kelly E J and Wishner R P 1965 Matched-filter theory for high-velocity, accelerating targets *IEEE Trans. Military Electr.* **9** 56

[49] Swick D A 1969 A review of wideband ambiguity functions *NRL Report 6994*

[50] Daubechies 1990 The wavelet transform, time-frequency localization, and signal analysis *IEEE Trans. Inform. Theory* **36** 961

[51] Kaiser G 1994 *A Friendly Guide to Wavelets* (Boston: Birkhauser)

[52] Grossmann A and Morlet J 1984 Decomposition of Hardy functions into square integrable wavelets of constant shape *SIAM J. Math. Anal.* **15** 723

# Appendix A

## Ill-Posed Problems

In this appendix we will apply some results from functional analysis to the explanation of ill-posedness. This exposition will be necessarily short and relies on the traditional definition/theorem approach—although the theorems are presented without proof (cf [1–6] for a more complete treatment).

### A.1 COMPACTNESS OF A SET AND COMPACT OPERATORS

Let $\langle u, v \rangle_T = \int_T u(t)v^*(t)\, dt$ denote the inner product over $T \subset \mathbb{R}$ and let $\mathcal{H}$, $\mathcal{H}_1$ and $\mathcal{H}_2$ denote Hilbert spaces with respect to the $L^2$-norm $\|u\|_T = \langle u, u \rangle_T^{1/2}$. If $\|u\|_T < \infty$ then $u$ is said to be square integrable over $T$. The space $L^2(T)$ of square integrable functions over $T$ is a Hilbert space and, if we consider $h \in L^2(\mathcal{K})$ and $\rho \in L^2(\mathcal{Y})$, then the scattering/imaging equation (3.4–3) can be written as

$$h = F\rho \tag{A-1}$$

where $F$ is the linear operator corresponding to $h(k) = \int_\mathcal{Y} f(k, y)\rho(y)\, dy$ which maps $\rho$ to the measurements $h$.

Let $\mathcal{H}_1 \subset \mathcal{H}$. The *closure* of $\mathcal{H}_1$, written $\overline{\mathcal{H}}_1$, is $\mathcal{H}_1$ together with all of its limit points. $\mathcal{H}_1$ is said to be *closed* if $\mathcal{H}_1 = \overline{\mathcal{H}}_1$. $\mathcal{H}_2 \subset \mathcal{H}$ is said to be *open* iff its complement is closed. These concepts are topological and introduced here so that we can define the important idea of *compactness* of a set.

*Definition* A.1. A system $\{S_j\}$ of open subsets of $\mathcal{H}$ is an *open covering* if each element in $\mathcal{H}$ belongs to at least one $S_j$. $C \subset \mathcal{H}$ is said to be *compact* if every open cover of $C$ contains a finite subcover.

Compactness of a set is also a topological property and can be somewhat difficult to visualize from this definition. A set consisting of a finite number of points is obviously compact, for example, but the definition is more general (and useful). The next theorem greatly simplifies the characterization of compact sets in $\mathbb{R}^n$.

*Theorem* A.1 *(Heine–Borel)*. The compact subsets of $\mathbb{R}^n$ are the closed bounded subsets of $\mathbb{R}^n$.

Our present interests concern the behavior of $F$ and, in particular, in solution algorithms for equation (4.1–2) and the behavior of these solutions. The basic problems are: these data are collected over a finite bandwidth $k \in [k_1, k_2]$; and these data will usually be contaminated by measurement errors ('noise'). This noise cannot generally be removed from these data and gets folded into the solution. We would like small amounts of noise (small variations in our data) to result in small variations in the solution. As we shall see, this is not usually what happens.

*Definition* A.2.    Let $F$ be a linear operator from $\mathcal{H}_1$ to $\mathcal{H}_2$. The *domain* $D(F)$ of $F$ is that subset of $\mathcal{H}_1$ over which $F$ is defined. The *range* $R(F)$ of $F$ is defined by $R(F) = \{h | h = F\rho \quad \text{for} \quad \rho \in D(F)\}$, and the *null space* $N(F)$ of $F$ is defined by $N(F) = \{\rho | F\rho = 0\}$. The *norm* of $F$ is defined by

$$\|F\| = \sup_{\substack{\rho \in D(F) \\ \rho \neq 0}} (\|F\rho\|_{\mathcal{K}} / \|\rho\|_{\mathcal{Y}}) \tag{A-2}$$

and $F$ is said to be *bounded* if it has finite norm.

*Definition* A.3.    A linear operator $F$ mapping $\mathcal{H}_1$ to $\mathcal{H}_2$ is called *compact* if $\overline{F(\mathcal{B})}$ is compact for every bounded subset $\mathcal{B}$ of $\mathcal{H}_1$.

Compactness of sets has been used to define compact operators, and compactness of operators, it turns out, can be used to define *continuity*. We have:

*Theorem* A.2.    Compact operators are continuous.

*Theorem* A.3.    The linear operator $F$, from $\mathcal{H}_1$ to $\mathcal{H}_2$, is bounded iff it is continuous.

In general, compactness of a particular operator can be awkward to establish from definition A.3. The following theorem offers a very useful alternative approach.

*Theorem* A.4.    Let $F$ be the linear operator defined by equation (4.1–2), and suppose that the kernel $f(\cdot, \cdot)$ is square integrable over $\mathcal{K} \times \mathcal{Y}$. Then $F$ is a compact operator from $L^2(\mathcal{Y})$ to $L^2(\mathcal{K})$. The converse is not generally true.

Compact operators are very important in many problems of mathematical physics. In addition to continuity (compact operators are sometimes referred to as 'completely continuous'), this importance is also due to the property of spectral decomposition.

## A.2  SINGULAR VALUE DECOMPOSITION

*Definition* A.4.  The *orthogonal complement* $S^\perp$ of a subset $S \subset \mathcal{H}$ is given by $S^\perp = \{u | \langle u, v \rangle = 0 \quad \text{for all} \quad v \in S\}$.

*Definition* A.5.  If $F$ is a bounded linear operator from $\mathcal{H}_1$ to $\mathcal{H}_2$, its *adjoint* $F^\dagger$ is the bounded linear operator from $\mathcal{H}_2$ to $\mathcal{H}_1$ satisfying $\langle F\rho, h \rangle_\mathcal{K} = \langle \rho, F^\dagger h \rangle_\mathcal{Y}$ for all $\rho \in \mathcal{H}_1$ and $h \in \mathcal{H}_2$.

*Theorem* A.5.  Let $F$ be a continuous linear operator from $\mathcal{H}_1$ to $\mathcal{H}_2$. Then $[R(F)]^\perp = N(F^\dagger)$ and $[N(F)]^\perp = \overline{R(F^\dagger)}$. Furthermore, $\|F^\dagger\| = \|F\|$.

*Theorem* A.6 *(Hilbert–Schmidt).*  Let $G$ be a compact self-adjoint $(G = G^\dagger)$ operator from $\mathcal{H}$ into itself with eigenvalues $\{\lambda_1, \lambda_2, \ldots\}$. Let $\{g_1, g_2, \ldots\}$ be the corresponding orthonormal eigenvectors (i.e. $Gg_j = \lambda_j g_j$). Then for any $\xi \in \mathcal{H}$, we have

$$G\xi = \sum_{j=1}^{\infty} \lambda_j \langle \xi, g_j \rangle g_j. \tag{A–3}$$

*Theorem* A.7.  The non-zero eigenvalues $\{\lambda_1, \lambda_2, \ldots\}$ of a compact, self-adjoint operator $G$ from $\mathcal{H}$ into itself are isolated and form a sequence that converges to zero (unless their number is finite).

*Definition* A.6.  Let $F$ be a compact linear operator from $\mathcal{H}_1$ to $\mathcal{H}_2$. Then $F^\dagger F$ is a compact linear operator mapping $\mathcal{H}_1$ to itself. The eigenvalues of $F^\dagger F$ are all non-negative and can be listed in order of decreasing magnitude so that $\lambda_1 \geqslant \lambda_2 \geqslant \ldots \geqslant 0$. Let $f_1, f_2, \ldots$ be a corresponding sequence of orthonormal eigenvectors of $F^\dagger F$ and define $\mu_j = 1/\sqrt{\lambda_j}$ and $h_j = \mu_j F f_j$. $\{h_j\}$ is an orthonormal sequence in $\mathcal{H}_2$ satisfying $\mu_j F^\dagger h_j = f_j$ and the sequence of triples $\{h_j, f_j; \mu_j\}$ is called a *singular system* for $F$.

*Theorem* A.8.  From definition A.6 we have: $\{h_j\}$ is an orthonormal basis for $\overline{R(F)} = [N(F^\dagger)]^\perp$ and $\{f_j\}$ is an orthonormal basis for $\overline{R(F^\dagger)} = [N(F)]^\perp$.

*Theorem* A.9.  Let $F$ be a compact linear operator mapping $\mathcal{H}_1$ into $\mathcal{H}_2$, and let $\{h_j, f_j; \mu_j\}$ be a singular system for $F$. The equation $h = F\rho$ has a solution if and only if the following two conditions hold:

(i)  $h \in [N(F^\dagger)]^\perp$,

(ii)  $\sum_{j=1}^{\infty} \mu_j^2 |\langle h, h_j \rangle|^2 < \infty$.

When these conditions are met, the solution is given by

$$\rho = \sum_{j=1}^{\infty} \mu_j \langle h, h_j \rangle f_j. \tag{A–4}$$

Because of equation (A–4), each measurement of the form of equation (4.1–2) may be considered to be the sum of projections of an unknown $\rho$ onto response vectors $f_j$. Only those components of $\rho$ that lie in the subspace spanned by the set of all $f_j$ contribute to the measurement. This subspace is called the *measurement space* and the components of $\rho$ in the subspace orthogonal to it (the null space) do not contribute to the measurements. Any $\rho$ lying wholly in the null space will yield zero measurements.

We can see from definition A.6 and theorem A.7 that the $\mu_j$ become arbitrarily large as $j$ increases (unless $F^{\dagger}F$ is finite-dimensional). If $F^{\dagger}F$ has only a finite number of eigenvalues then the sum in equation (A–3) will be finite and $F^{\dagger}F$ is said to be of finite rank. In this case, $R(F^{\dagger}F)$ is a finite-dimensional set. Practicable measurements, of course, will be finite and $\mu_j$ cannot actually become infinite. Note, however, that increasing the dimension of the measurement space (by making more measurements) will generally *increase* the maximum value of $\mu_j$.

## A.3   LEAST-SQUARES SOLUTIONS AND ILL-POSEDNESS

When the *actual* measurements are of the form

$$h = F\rho + n \qquad\qquad (A–5)$$

where $n \in L^2(\mathcal{K})$ is an unknown noise component that will not generally be restricted to $\overline{R(F)}$, then the model equation (A–1) may be inconsistent with the data. The idea behind a 'least-squares' solution to equation (A–5) is to find the $\rho$ for which $F\rho$ lies closest to the measured data.

*Theorem* A.10.   Let $F$ be a compact linear operator from $\mathcal{H}_1$ to $\mathcal{H}_2$. The following conditions are equivalent:

  (i)  $\|Ff_{\mathrm{ls}} - h\|_{\mathcal{K}} = \inf_{\rho \in \mathcal{H}_1} \|F\rho - h\|_{\mathcal{K}}$. We call this $f_{\mathrm{ls}}$ the *least-squares* solution.

  (ii)  $F^{\dagger}Ff_{\mathrm{ls}} = F^{\dagger}h$. This is the *normal equation*.

  (iii)  $Ff_{\mathrm{ls}} = Ph$, where $P$ is the projection operator mapping $\mathcal{H}_2$ onto $\overline{R(F)}$.

It follows from theorem A.10-(iii) that a least-squares solution exists iff $h \in R(F) + [R(F)]^{\perp}$. When this condition is satisfied, there is a unique least-squares solution with minimum norm, denoted by $F^{-}h$.

*Definition* A.7.   The operator $F^{-}$ maps $R(F) + [R(F)]^{\perp}$ into $\mathcal{H}_1$ and is called the *pseudoinverse* of $F$. We have $N(F^{-}) = [R(F)]^{\perp}$ and $R(F^{-}) = [N(F)]^{\perp}$.

*Theorem* A.11.   Let $F$ be a compact linear operator from $\mathcal{H}_1$ to $\mathcal{H}_2$. Then $F^{-}$ is bounded iff $F$ has finite rank.

*Theorem* A.12.    Let $F$ be a compact linear operator from $\mathcal{H}_1$ to $\mathcal{H}_2$ with singular system $\{h_j, f_j; \mu_j\}$, and assume that $h \in R(F) + [R(F)]^\perp$ with $Ph \in R(F)$. Then

$$F^- h = \sum_{j=1}^{\infty} \mu_j \langle Ph, h_j \rangle f_j = \sum_{j=1}^{\infty} \mu_j \langle h, h_j \rangle f_j. \qquad \text{(A--6)}$$

In summary, we have shown that when $F$ is compact, $\mu_j \to \infty$ (definition A.6 and theorem A.7) and, consequently, perturbations in higher modes can be greatly amplified. In particular, this result together with theorem A.4, means that $h = F\rho$ will generally be ill-posed when $f(k, y)$ is square-integrable and non-degenerate.

# REFERENCES

[1]    Engl H W, Hanke M and Neubauer A 1996 *Regularization of Inverse Problems (Mathematics and Its Applications Series)* vol 375 (New York: Kluwer Academic)

[2]    Rushforth C K 1987 Signal restoration, functional analysis, and Fredholm integral equations of the first kind *Image Recovery: Theory and Application* ed H Stark (Orlando, FL: Academic)

[3]    Nashed M Z 1976 On moment-discretization and least-squares solutions of linear integral equations of the first kind *J. Math. Anal. Appl.* **53** 359

[4]    Nashed M Z 1981 Operator-theoretic and computational approaches to ill-posed problems with applications to antenna theory *IEEE Trans. Antennas Propag.* **29** 220

[5]    Andrews H C and Hunt B R 1977 *Digital Image Restoration* (Englewood Cliffs, NJ: Prentice-Hall)

[6]    Sage A P and Melsa J L 1979 *Estimation Theory with Applications to Communications and Control* (New York: Krieger)

# Bibliography

Adachi S and Uno T 1993 One-dimensional target profiling by electromagnetic backscattering *J. Electromag. Waves Appl.* **7** 403

Andrews H C and Hunt B R 1977 *Digital Image Restoration* (Englewood Cliffs, NJ: Prentice-Hall)

Asseo S J 1974 Effect of monopulse thresholding on tracking multiple targets *IEEE Trans. Aerospace Electr. Syst.* **10** 504

Ausherman D A, Kozma A, Walker J L, Jones H M and Poggio E C 1984 Developments in radar imaging *IEEE Trans. Aerospace Electr. Syst.* **20** 363

Baltes H P 1980 *Inverse Scattering Problems in Optics, Topics in Current Physics* ed H P Baltes (New York: Springer)

Baum C E 1971 On the singularity expansion method for the solution of electromagnetic interaction problems *Interaction Note 88* (Kirtland AFB, NM: Air Force Weapons Laboratory)

Baum C E 1976 The singularity expansion method *Transient Electromagnetic Fields* ed L B Felsen (New York: Springer)

Baum C E 1997 Discrimination of buried targets via the singularity expansion *Inverse Problems* **13** 557

Baum C E, Rothwell E J, Chen K-M and Nyquist D P 1991 The singularity expansion method and its application to target identification *Proc. IEEE* **79** 1481

Baynard J-P R and Schaubert D H 1990 Target identification using optimization techniques *IEEE Trans. Antennas Propag.* **38** 450

Bell M R and Grubbs R A 1993 JEM modeling and measurement for radar target identification *IEEE Trans. Aerospace Electr. Syst.* **29** 73

Bennett C L 1981 Time domain inverse scattering *IEEE Trans. Antennas Propag.* **29** 213

Bennett C L, Auckenthaler A M, Smith R S and DeLorenzo J D 1973 Space time integral equation approach to the large body scattering problem *RADC-CR-73-70 AD763794* (Sudbury, MA: Sperry Rand Research Center)

Berizzi F and Corsini G 1996 Autofocusing of inverse synthetic aperture radar images using contrast optimization *IEEE Trans. Aerospace Electr. Syst.* **32** 1185

Berkowitz R S *Modern Radar* (New York: Wiley)

Bernfeld M 1984 Chirp Doppler radar *Proc. IEEE Lett.* **72** 540

Berni A J 1975 Target identification by natural resonance estimation *IEEE Trans. Aerospace Electr. Syst.* **11** 147

Bleistein N 1989 Large wave number aperture-limited Fourier inversion and inverse scattering *Wave Motion* **11** 113

Bleistein N and Handelsman R A 1975 *Asymptotic Expansions of Integrals* (New York: Holt, Rinehart and Winston)

Bocker R P and Jones S 1992 ISAR motion compensation using the burst derivative measure as a focal quality indicator *Int. J. Imaging Syst. Tech.* **4** 286

136

Boerner W-M 1980 Polarization utilization in electromagnetic inverse scattering *Inverse Scattering Problems in Optics* ed H P Baltes (New York: Springer) ch 7

Boerner W-M 1985 *Inverse Methods in Electromagnetic Imaging (NATO ASI Series C: Mathematical and Physical Sciences)* vol 143, ed W-M Boerner *et al* (Dordrecht: Reidel)

Bojarski N N 1967 Three dimensional electromagnetic short pulse inverse scattering *Special Projects Lab. Report* (Syracuse, NY: Syracuse University Research Corporation)

Bojarski N N 1982 A survey of the physical optics inverse scattering identity *IEEE Trans. Antennas Propag.* **30** 980

Borden B 1986 High-frequency statistical classification of complex targets using severely aspect-limited data *IEEE Trans. Antennas Propag.* **34** 1455

Borden B 1991 Diversity methods in phase monopulse tracking—a new approach *IEEE Trans. Aerospace Electr. Syst.* **27** 877

Borden B 1992 Phase monopulse tracking and its relationship to non-cooperative target recognition *IMA Volumes in Mathematics and its Applications, Radar and Sonar, Part 2* vol 39, ed F A Grünbaum, M Bernfeld and R E Blahut (New York: Springer)

Borden B 1994 Problems in airborne radar target recognition *Inverse Problems* **10** 1009

Borden B 1994 Requirements for optimal glint reduction by diversity methods *IEEE Trans. Aerospace Electr. Syst.* **30** 1108

Borden B 1995 Enhanced range profiles for radar-based target classification using mono-pulse tracking statistics *IEEE Trans. Antennas Propag.* **43** 759

Borden B 1997 Some issues in inverse synthetic aperture radar image reconstruction *Inverse Problems* **13** 571

Borden B 1997 An observation about radar imaging of re-entrant structures with implications for automatic target recognition *Inverse Problems* **13** 1441

Bowen E G 1987 *Radar Days* (Bristol: Adam Hilger)

Brittingham J N, Miller E K and Willows J L 1980 Pole extraction from real frequency information *Proc. IEEE* **68** 263

Brown W M and Fredricks R J 1969 Range-Doppler imaging with motion through resolution cells *IEEE Trans. Aerospace Electr. Syst.* **5** 98

Brown W M and Porcello L J 1969 An introduction to synthetic aperture radar *IEEE Spectrum* **6** 52

Burdic W S 1968 *Radar Signal Analysis* (Englewood Cliffs, NJ: Prentice-Hall)

Cafforio C, Prati C and Rocca E 1991 SAR data focusing using seismic migration techniques *IEEE Trans. Aerospace Electr. Syst.* **27** 194

Carin L, Felsen L B, Kralj D, Pillai S U and Lee W C 1994 Dispersive modes in the time domain: analysis and time-frequency representation *IEEE Microwave Guided Wave Lett.* **4** 23

Cartwright M L 1930 The zeros of certain integral functions *Q. J. Math., Oxford Ser. (1)* **1** 38

Chamberlain N F, Walton E K and Garber F D 1991 Radar target identification of aircraft using polarization-diverse features *IEEE Trans. Aerospace Electr. Syst.* **27** 58

Chen C C and Andrews H C 1980 Multifrequency imaging of radar turntable data *IEEE Trans. Aerospace Electr. Syst.* **16** 15

Chen C-C and Andrews H C 1980 Target-motion-induced radar imaging *IEEE Trans. Aerospace Electr. Syst.* **16** 2

Chen K-M, Nyquist D P, Rothwell E J, Webb L L and Drachman B 1986 Radar target discrimination by convolution of radar returns with extinction pulses and single-mode extraction signals *IEEE Trans. Antennas Propag.* **34** 896

Chuang C W and Moffatt D L 1976 Natural resonance of radar target via Prony's method and target discrimination *IEEE Trans. Aerospace Electr. Syst.* **12** 583

Cloude S R and Pottier E 1996 A review of target decomposition theorems in radar polarimetry *IEEE Trans. Geosci. Remote Sensing* **34** 498

Cohen L 1989 Time-frequency distributions - a review *Proc. IEEE* **77** 941

Collot G 1991 Fixed/rotary wings classification recognition *Proc. CIE 1991 Int. Conf. on Radar (CICR-91) (Beijing, China)* (New York: IEEE) p 610

Colton D and Kress R 1983 *Integral Equation Methods in Scattering Theory* (New York: Wiley)

Cook C E and Bernfeld M 1967 *Radar Signals* (New York: Academic)

Copeland J R 1960 Radar target classification by polarization properties *Proc. IRE* **48** 1290

Das Y and Boerner W-M 1978 On radar target shape estimation using algorithms for reconstruction from projections *IEEE Trans. Antennas Propag.* **26** 274

Daubechies 1990 The wavelet transform, time-frequency localization, and signal analysis *IEEE Trans. Inform. Theory* **36** 961

DeGraaf S R 1988 Parametric estimation of complex 2-d sinusoids *IEEE Fourth Ann. ASSP Workshop on Spectrum Estimation and Modeling* (New York: IEEE) p 391

Delano R H 1953 A theory of target glint or angular scintillation in radar tracking *Proc. IRE* **41** 1778

Delisle G Y and Wu H 1994 Moving target imaging and trajectory computation using ISAR *IEEE Trans. Aerospace Electr. Syst.* **30** 887

Deschamps G A 1951 Geometrical representation of the polarization of a plane electromagnetic wave *Proc. IRE* **39** 540

Dunn J H and Howard D D 1959 The effects of automatic gain control performance on the tracking accuracy of monopulse radar systems *Proc. IRE* **47** 430

Dural G and Moffatt D L 1994 ISAR imaging to identify basic scattering mechanisms *IEEE Trans. Antennas Propag.* **42** 99

Engl H W, Hanke M and Neubauer A 1996 *Regularization of Inverse Problems (Mathematics and Its Applications Series)* vol 375 (New York: Kluwer Academic)

Erdélyi A, Magnus W, Oberhettinger F and Tricomi FG 1954 *Tables of Integral Transforms* vol II (New York: McGraw-Hill)

Feig E and Grünbaum A 1986 Tomographic methods in range-Doppler radar *Inverse Problems* **2**

Felsen L B 1987 Theoretical aspects of target classification *AGARD Lecture Series* no 152, ed L B Felsen (Munich: AGARD)

Fisher D E 1988 *A Race on the Edge of Time: Radar—the Decisive Weapon of World War II* (New York: McGraw-Hill)

Flammer C 1957 *Spheroidal Wave Functions* (Stanford, CA: Stanford University Press)

Frieden B R and Bajkova A T 1994 Bayesian cross-entropy of complex images *Appl. Opt.* **33** 219

Gardner W 1990 *Introduction to Random Processes* 2nd edn (New York: McGraw-Hill)

Gerchberg R W 1979 Super-resolution through error energy reduction *Opt. Acta* **14** 709

Gross F B and Young J D 1981 Physical optics imaging with limited aperture data *IEEE Trans. Antennas Propag.* **29** 332

Grossmann A and Morlet J 1984 Decomposition of Hardy functions into square integrable wavelets of constant shape *SIAM J. Math. Anal.* **15** 723

Gubonin N S 1965 Fluctuations of the phase front of the wave reflected from a complex target *Radio Eng. Electr. Phys.* **10** 718

Harris F J 1978 On the use of windows for harmonic analysis with the discrete Fourier transform *Proc. IEEE* **66** 1

Herman G T and Lent A 1976 A computer implementation of a Bayesian analysis of image reconstruction *Inform. Contr.* **31** 364

Hovenessian S A 1980 *Introduction to Synthesis Array and Imaging Radar* (Dedham, MA: Artech House)

Howard D D 1975 High range-resolution monopulse tracking radar *IEEE Trans. Aerospace Electr. Syst.* **11** 749

Hua Y, Baqai F and Zhu Y 1993 Imaging of point scatterers from step-frequency ISAR data *IEEE Trans. Aerospace Electr. Syst.* **29** 195

Huang C-C 1983 Simple formula for the RCS of a finite hollow circular cylinder *Electron. Lett.* **19** 854

Hudson S and Psaltis D 1993 Correlation filters for aircraft identification from radar range profiles *IEEE Trans. Aerospace Electr. Syst.* **29** 741

Hunt B R 1977 Bayesian methods in nonlinear digital image restoration *IEEE Trans. Comput.* **26** 219

Hutson V and Pym J S 1980 *Applications of Functional Analysis and Operator Theory* (New York: Academic)

Huynen J R 1978 Phenomenological theory of radar targets *Electromagnetic Scattering* ed P L E Uslenghi (New York: Academic)

Ishimaru A 1978 *Wave Propagation and Scattering in Random Media* (New York: Academic)

Jain A and Patel I 1990 Simulations of ISAR image errors *IEEE Trans. Instrum. Meas.* **39** 212

Johnson T W and Moffatt D L 1982 Electromagnetic scattering by open circular waveguides *Radio Sci.* **17** 1547

Kachelmyer A L 1992 Inverse synthetic aperture radar image processing *Laser Radar* **7** 193

Kaiser G 1994 *A Friendly Guide to Wavelets* (Boston: Birkhauser)

Kalnins E G and Miller W Jr 1992 A note on group contractions and radar ambiguity functions *IMA Volumes in Mathematics and its Applications, Radar and Sonar, Part 2* vol 39, ed F A Grünbaum, M Bernfeld and R E Blahut (New York: Springer) p 71

Keller J B 1959 The inverse scattering problem in geometrical optics and the design of reflectors *IRE Trans. Antennas Propag.* **7** 146

Kelly E J and Wishner R P 1965 Matched-filter theory for high-velocity, accelerating targets *IEEE Trans. Military Electr.* **9** 56

Kennaugh E M 1952 Polarization properties of target reflection *Technical Report 389–2* (Griffis AFB, NY)

Kennaugh E M 1981 The K-pulse concept *IEEE Trans. Antennas Propag.* **29** 327

Kennaugh E M and Moffatt D L 1965 Transient and impulse response approximations *Proc. IEEE* **53** 893

Kim H and Ling H 1993 Wavelet analysis of radar echo from finite-size targets *IEEE Trans. Antennas Propag.* **41** 200

Kirk J C Jr 1975 Motion compensation for synthetic aperture radar *IEEE Trans. Aerospace Electr. Syst.* **11** 338

Kleinman R E and van den Berg P M 1991 Iterative methods for solving integral equations *Radio Sci.* **26** 175

Knott E F and Senior T B A 1972 Cross polarization diagnostics *IEEE Trans. Antennas Propag.* **20** 223

Kong K K and Edwards J A 1995 Polar format blurring in ISAR imaging *Electron. Lett.* **31** 1502

Kraus J D 1988 *Antennas* (New York: McGraw-Hill)

Ksienski D A 1985 Pole and residue extraction from measured data in the frequency domain using multiple data sets *Radio Sci.* **20** 13

Langenberg K J, Brandfass M, Mayer K, Kreutter T, Brüll A, Fellinger P and Huo D 1993 Principles of microwave imaging and inverse scattering *EARSeL Adv. Remote Sensing* **2** 163

Leonov A I, Fomichev K I, Barton W F and Barton D 1986 *Monopulse Radar* (Dedham, MA: Artech House)

Levanon N 1988 *Radar Principles* (New York: Wiley)

Lewis R M 1969 Physical optics inverse diffraction *IEEE Trans. Antennas Propag.* **17** 308: correction 1970 *IEEE Trans. Antennas Propag.* **18** 194

Lind G 1968 Reduction of radar tracking errors with frequency agility *IEEE Trans. Aerospace Electr. Syst.* **4** 410

Lind G 1972 A simple approximate formula for glint improvement with frequency agility *IEEE Trans. Aerospace Electr. Syst.* **8** 854

Ling H, Chou R-C and Lee S W 1989 Shooting and bouncing rays: calculating the RCS of an arbitrary cavity *IEEE Trans. Antennas Propag.* **37** 194

Loomis J M, III and Graf E R 1974 Frequency-agility processing to reduce radar glint pointing error *IEEE Trans. Aerospace Electr. Syst.* **10** 811

Maass P 1992 Wideband approximation and wavelet transform *IMA Volumes in Mathematics and its Applications, Radar and Sonar, Part 2* vol 39, ed F A Grünbaum, M Bernfeld and R E Blahut (New York: Springer) p 83

Mager R D and Bleistein N 1978 An examination of the limited aperture problem of physical optics inverse scattering *IEEE Trans. Antennas Propag.* **26** 695

Maitre H 1981 Iterative superresolution: some new fast methods *Opt. Acta* **28** 973

Mandel L 1974 Interpretation of instantaneous frequency *Am. J. Phys.* **42** 840

Manson A C and Boerner W-M 1985 Interpretation of high-resolution polarimetric radar target down-range signatures using Kennaugh's and Huynen's target characteristic operator theories *Inverse Methods in Electromagnetic Imaging* ed W-M Boerner (Dordrecht: Reidel)

Marin L 1973 Natural-mode representation of transient scattered fields *IEEE Trans. Antennas Propag.* **21** 809

Marin L and Latham R W 1972 Representation of transient scattered fields in terms of free oscillations of bodies *Proc. IEEE* **60** 640

Mensa D, Heidbreder G and Wade G 1980 Aperture synthesis by object rotation in coherent imaging *IEEE Trans. Nucl. Sci.* **27** 989

Mensa D L 1981 *High Resolution Radar Imaging* (Dedham, MA: Artech House)

Mensa D L, Halevy S and Wade G 1983 Coherent Doppler tomography for microwave imaging *Proc. IEEE* **71** 254

Merserau R M and Oppenheim A V 1974 Digital reconstruction of multidimensional signals from their projections *Proc. IEEE* **62** 1319

Mevel J 1976 Procedure de reconnaissance des formes l'ide d'un radar monostatique *Ann. Telecom.* **131** 111

Michalski K A 1981 Bibliography of the singularity expansion method and related topics *Electromagnetics* **1** 493

Michalski K A 1982 On the class 1 coupling coefficient performance in the SEM expansions for current density on a scattering object *Electromagnetics* **2** 201

Moffatt D L and Mains R K 1975 Detection and discrimination of radar targets *IEEE Trans. Antennas Propag.* **23** 358

Moghaddar A and Walton E K 1993 Time-frequency distribution analysis of scattering from waveguides *IEEE Trans. Antennas Propag.* **41** 677

Mohammad-Djafari A and Demoment G 1989 Maximum entropy Fourier synthesis with application to diffraction tomography *Appl. Opt.* **26** 1745

Moll J W and Seecamp R G 1969 Calculation of radar reflecting properties of jet engine intakes using a waveguide model *IEEE Trans. Aerospace Electr. Syst.* **6** 675

Moore J and Ling H 1995 Super-resolved time-frequency analysis of wideband backscattered data *IEEE Trans. Antennas Propag.* **43** 623

Morgan M A 1984 Singularity expansion representations of fields and currents in transient scattering *IEEE Trans. Antennas Propag.* **32** 466

Morgan M A 1988 Scatterer discrimination based upon natural resonance annihilation *J. Electromag. Waves Appl.* **2** 481

Muchmore R B 1960 Aircraft scintillation spectra *IRE Trans. Antennas Propag.* **8** 201

Müller C 1969 *Foundations of the Mathematical Theory of Electromagnetic Waves* (New York: Springer)

Munson D C, O'Brien D and Jenkins W K 1983 A tomographic formulation of spotlight-mode synthetic aperture radar *Proc. IEEE* **71** 917

Naparst H 1991 Dense target signal processing *IEEE Trans. Inform. Theory* **37** 317

Nashed M Z 1976 On moment-discretization and least-squares solutions of linear integral equations of the first kind *J. Math. Anal. Appl.* **53** 359

Nashed M Z 1981 Operator-theoretic and computational approaches to ill-posed problems with applications to antenna theory *IEEE Trans. Antennas Propag.* **29** 220

Nichols L A 1975 Reduction of radar glint for complex targets by use of frequency agility *IEEE Trans. Aerospace Electr. Syst.* **11** 647

Noel B 1991 *Ultra-Wideband Radar, Proc. First Los Alamos Symp.* ed B Noel (New York: CRC Press)

North D O 1943 An analysis of the factors which determine signal-noise discrimination in pulsed carrier systems *RCA Lab. Report PTR-6C* (New York: IEEE)

Nullin C M and Aas B 1987 Experimental investigation of correlation between fading and glint for aircraft targets *Proc. Radar-87 Conf.* (New York: IEEE) p 540

Nuttall A H 1966 On the quadrature approximation to the Hilbert transform of modulated signals *Proc. IEEE* **54** 1458

Papoulis A 1965 *Probability, Random Variables, and Stochastic Processes* (New York: McGraw-Hill)

Papoulis A 1975 A new algorithm in spectral analysis and band-limited extrapolation *IEEE Trans. Circuits Syst.* **22** 735

Park S-W and Cordaro J T 1988 Improved estimation of SEM parameters from multiple observations *IEEE Trans. Aerospace Electr. Syst.* **30** 145

Pathak P H and Burkholder R J 1991 High-frequency electromagnetic scattering by open-ended waveguide cavities *Radio Sci.* **26** 211

Pathak P H, Chuang C W and Liang M C 1986 Inlet modeling studies *Ohio State University, ElectroScience Lab Report 717674-1*

Pearson L W 1983 Present thinking on the use of the singularity expansion in electromagnetic scattering computation *Wave Motion* **5** 355

Perry W L 1974 On the Bojarski–Lewis inverse scattering method *IEEE Trans. Antennas Propag.* **22** 826

Pham D T 1998 Applications of unsupervised clustering algorithms to aircraft identification using high range resolution radar *Non-Cooperative Air Target Identification Using Radar, AGARD Symp. (22–24 April, Mannheim, Germany)* (Quebec: Canada Communication group Inc) p 16

Poggio A J and Miller E K 1973 Integral equation solutions to three dimensional scattering problems *Computer Techniques for Electromagnetics* ed R Mittra (London: Pergamon)

Prickett M J and Chen CC 1980 Principles of inverse synthetic aperture radar (ISAR) imaging *IEEE 1980 EASCON Record* p 340

Rhebergen J B, van den Berg P M and Habashy T M 1997 Iterative reconstruction of images from incomplete spectral data *Inverse Problems* **13** 829

Rhodes D R 1959 *Introduction to Monopulse* (New York: McGraw-Hill)

Rihaczek A W 1969 *Principles of High-Resolution Radar* (New York: McGraw-Hill)

Rihaczek A W and Hershkowitz S J 1996 *Radar Resolution and Complex-Image Analysis* (Dedham MA: Artech House)

Rooney P G 1980 On the $\mathcal{Y}_\nu$ and $\mathcal{H}_\nu$ transformations *Can. J. Math.* **32** 1021

Rosenbaum-Raz S 1976 On scatterer reconstruction from far-field data *IEEE Trans. Antennas Propag.* **24** 66

Ross D C, Volakis J L and Hristos T A 1995 Hybrid finite element analysis of jet engine inlet scattering *IEEE Trans. Antennas Propag.* **43** 277

Rushforth C K 1987 Signal restoration, functional analysis, and Fredholm integral equations of the first kind *Image Recovery: Theory and Application* ed H Stark (Orlando, FL: Academic)

Sage A P and Melsa J L 1979 *Estimation Theory with Applications to Communications and Control* (New York: Krieger)

Sanz J L C and Huang T S 1983 Unified Hilbert space approach to iterative least-squares linear signal restoration *J. Opt. Soc. Am.* **73** 1455

Sherman S M 1984 *Monopulse Principles and Techniques* (Dedham, MA: Artech House)

Sims R J and Graf E R 1971 The reduction of radar glint by diversity techniques *IEEE Trans. Antennas Propag.* **19** 462

Skolnik M L 1980 *Introduction to Radar Systems* 2nd edn (New York: McGraw-Hill)

Slepian D and Pollak H O 1961 Prolate spheroidal wave functions, Fourier analysis and uncertainty – I *Bell Syst. Tech. J.* **40** 43

Stark H 1987 *Image Recovery: Theory and Practice* ed H Stark (Orlando, FL: Academic)

Steinberg B D 1988 Microwave imaging of aircraft *IEEE Proc.* **76** 1578

Stoker J J 1969 *Differential Geometry* (New York: Wiley-Interscience)

Stratton J A 1941 *Electromagnetic Theory* (New York: McGraw-Hill)

Swick D A 1969 A review of wideband ambiguity functions *NRL Report 6994*

Swords S S 1986 *Technical History of the Beginnings of Radar* (London: Peregrinus)

Tabbara W 1973 On an inverse scattering method *IEEE Trans. Antennas Propag.* **21** 245

Tabbara W 1975 On the feasibility of an inverse scattering method *IEEE Trans. Antennas Propag.* **23** 446; correction 1977 *IEEE Trans. Antennas Propag.* **25** 286

Titchmarsh E C 1925 The zeros of certain integral functions *Proc. Lond. Math. Soc.* **25** 283

Trintinalia L C and Ling H 1996 Extraction of waveguide scattering features using joint time-frequency ISAR *IEEE Microwave Guided Wave Lett.* **6** 10

Trintinalia L C and Ling H 1997 Joint time-frequency ISAR using adaptive processing *IEEE Trans. Antennas Propag.* **45** 221

Trischman J, Jones S, Bloomfield R, Nelson E and Dinger R 1994 An X-band linear frequency modulated radar for dynamic aircraft measurement *AMTA Proc. (Long Beach, CA)* (New York: AMTA) p 431

Trischman J A 1996 Real-time motion compensation algorithms for ISAR imaging of aircraft *Proc. SPIE*

Tsao J and Steinberg B D 1988 Reduction of sidelobe and speckle artifacts in microwave imaging: the CLEAN technique *IEEE Trans. Antennas Propag.* **36** 543

US Air Force 1974 *Radar Signature Measurements of BQM-34A and BQM-34F Target Drones* ASFWC-TR-74-01, AD785219 (Holloman AFB, NM: The Radar Target Scattering Division, 6585th Test Group)

VanBlaricum M L and Mittra R 1975 A technique for extracting the poles and residues of a system directly from its transient response *IEEE Trans. Antennas Propag.* **23** 777

VanBlaricum M L and Mittra R 1978 Problems and solutions associated with Prony's method for processing transient data *IEEE Trans. Antennas Propag.* **26** 174

Walker J L 1980 Range-Doppler imaging of rotating objects *IEEE Trans. Aerospace Electr. Syst.* **16** 23

Walsh T E 1978 Military radar systems: history, current position and future forecast *Microwave J.* **21** 87

Wang T M and Ling H 1991 Electromagnetic scattering from three-dimensional cavities via a connection scheme *IEEE Trans. Antennas Propag.* **39** 1501

Watanabe Y, Itoh T and Sueda H 1996 Motion compensation for ISAR via centroid tracking *IEEE Trans. Aerospace Electr. Syst.* **32** 1191

Wehner D R 1987 *High Resolution Radar* (Norwood, MA: Artech House)

Wehner D R, Prickett M J, Rock R G and Chen C C 1979 Stepped frequency radar target imagery, theoretical concept and preliminary results *Technical Report 490* (San Diego, CA: Naval Ocean Systems Command)

Weiss M R 1968 Inverse scattering in the geometric-optics limit *J. Opt. Soc. Am.* **58** 1524

Witt H R and Price E L 1968 Scattering from hollow conducting cylinders *Proc. Inst. Electr. Eng.* **115** 94

Woodward P M 1953 *Probability and Information Theory* (New York: Pergamon)

Wu H and Delisle G Y 1996 Precision tracking algorithms for ISAR imaging *IEEE Trans. Aerospace Electr. Syst.* **32** 243

Wu H, Grenier D and Fang D 1995 Translational motion compensation in ISAR processing *IEEE Trans. Image Proc.* **4** 1561

# Index

Printed in the United States
By Bookmasters